A Rehabilitated Estuarine Ecosystem

A Rehabilitated Estuarine Ecosystem

The environment and ecology of the Thames Estuary

Edited by Martin J. Attrill
Department of Biological Sciences,
University of Plymouth,
Plymouth, UK

SPRINGER SCIENCE+BUSINESS MEDIA, LLC

Library of Congress Cataloging in Publication Card Number: 98-70273

ISBN 978-1-4613-4671-5 ISBN 978-1-4419-8708-2 (eBook)
DOI 10.1007/978-1-4419-8708-2

Originally published by Kluwer Academic Publishers in 1998
Softcover reprint of the hardcover 1st edition 1998

Contents

Contributors xi
Acknowledgements xiv

1 Introduction 1
Jim Green
1.1 The Thames estuary: a personal view 1
1.2 Cleaning up the Thames and the role of this book 2

2 The Thames estuary: a history of the impact of humans on the environment and a description of the current approach to environmental management 5
Derek Tinsley
2.1 Introduction 5
2.2 Definition of the study area 6
2.3 Basic description of the Thames estuary 7
2.3.1 Width and depth 7
2.3.2 Freshwater inputs 8
2.3.3 Current use of the estuary 8
2.3.4 Estuary habitats 9
2.4 The Thames estuary before 1800 9
2.4.1 Methodology 9
2.4.2 The estuary environment before 1800 10
2.4.3 The presence and abundance of estuarine organisms before 1800 10
2.5 A history of the impact of humans on the environment (1800 to the present day) 11
2.5.1 Water quality 11
2.5.2 The decline in water quality in the nineteenth century 11
2.5.3 Interceptor sewers for London and chemical treatment of the sewage 12
2.5.4 The decline in water quality in the twentieth century 13
2.5.5 The Pippard Report and the Water Pollution Research Laboratory study 14

2.5.6 The Royal Commission on Environmental Pollution (Third Report) 15
2.6 Water quality management during the Thames Water Authority era 16
2.6.1 Water quality objectives 16
2.6.2 A pollution budget for dissolved oxygen 18
2.6.3 Improved water quality resulting from better sewage treatment 19
2.6.4 The threat to water quality from storm discharges 19
2.6.5 Control of toxic, non-biodegradable substances 20
2.7 Habitat loss as a result of flood defence work and land-claim 20
2.8 The current approach to environmental management 21
2.8.1 Water quality management 21
2.8.2 Operating agreements between the EA, Thames Region and Thames Water 25
2.8.3 The current approach to habitat management 25

3 Physical and chemical characteristics 27
Jane Kinniburgh
3.1 Introduction 27
3.2 Physical characteristics 27
3.2.1 Topography 27
3.2.2 Water levels 28
3.2.3 Water movements 30
3.2.4 Freshwater flow 30
3.2.5 Temperature 33
3.2.6 Suspended solids 35
3.3 Chemical characteristics 37
3.3.1 Salinity 37
3.3.2 Dissolved oxygen 40
3.3.3 Nutrients 45
3.3.4 Dangerous substances 46
3.4 Modelling47
3.5 Summary 48

4 The algae of the Thames estuary: a reappraisal 49
Ian Tittley and David John
4.1 Introduction 49
4.1.1 Backgound 49
4.1.2 Benthic algal habitats in the tidal estuary 50

4.2 The present reappraisal 51
4.2.1 Method 51
4.3 The algal flora by habitat 53
4.3.1 Marine to brackish-water reaches 53
4.3.2 Freshwater tidal reach 54
4.4 Seasonality 55
4.5 Changes in occurrence and distribution 57
4.6 Distribution along the estuary 58
4.7 Future studies 64
4.8 Conclusion 65

5 **The zooplankton communities of the Thames estuary** 67
Chris Gordon, Anthony Bark and Roland Bailey
5.1 Introduction 67
5.2 The current study on the zooplankton of the Thames estuary 68
5.3 Composition of the zooplankton 70
5.3.1 Protozoa, Cnidaria, Ctenophora and Rotifera 70
5.3.2 Cladocera 71
5.3.3 Copepoda 71
5.3.4 Mysidacea 71
5.3.5 Amphipoda 72
5.4 Spatial and temporal distributions of the major groups 72
5.4.1 *Eurytemora affinis* 72
5.4.2 *Gammarus zaddachi* 73
5.4.3 *Neomysis integer* 73
5.4.4 Cyclopoids 75
5.5 Changes in dominance of major species 75
5.6 Vertical and lateral distributions 79
5.7 Comparison with zooplankton communities in other estuaries 81
5.8 Zooplankton functioning 82
5.8.1 Zooplankton feeding 82
5.8.2 Zooplankton as food organisms 83
5.8.3 Other observations 84
5.9 Conclusions 84

6 **The benthic macroinvertebrate communities of the Thames estuary** 85
Martin Attrill
6.1 Introduction 85
6.2 Thames Estuary Benthic Programme 86
6.3 The Thames divided 86

6.4 Benthic macroinvertebrates – zone by zone 88
6.4.1 Zone 1: Teddington to London Bridge 88
6.4.2 Zone 2: London Bridge to Crossness 94
6.4.3 Zone 3: Crossness to Gravesend 98
6.4.4 Zone 4: Outer Estuary I 101
6.4.5 Zone 5: Outer Estuary II 110
6.5 Conclusions 112

7 Temporal changes in the movements and abundance of Thames estuary fish populations 115
Myles Thomas
7.1 Introduction 115
7.2 Sampling method 116
7.3 Species numbers 119
7.4 Diversity 121
7.5 Species abundance 122
7.5.1 Flounder 122
7.5.2 Sole 124
7.5.3 Whiting 124
7.5.4 Bass 125
7.5.5 Sand goby 126
7.5.6 Herring 126
7.5.7 Smelt 128
7.5.8 Salmon 128
7.6 Community composition 129
7.6.1 Species dendrogram 132
7.7 Trawling method 132
7.7.1 Method 134
7.7.2 Results 136
7.8 Summary 138

8 Host–parasite interactions: case studies of parasitic infections in migratory fish 141
Margaret Munro, Phillip Whitfield and Steve Lee
8.1 Introduction 141
8.2 The intestinal fish helminth *Pomphorhynchus laevis* 143
8.2.1 Introduction 143
8.2.2 Infection patterns in the intermediate host 145
8.2.3 The range of definitive host species 148
8.2.4 Infection patterns in flounder 150
8.2.5 A comparison of infections in flounder and smelt 152
8.2.6 Pathogenicity to estuarine fish 155
8.3 Virus-associated spawning papillomatosis in smelt 157

8.3.1 Introduction 157
8.3.2 Morphology and ultrastructure 158
8.3.3 Epidemiology 159
8.4 Other parasites of migratory fish 163
8.5 Summary and conclusions 165

9 The Thames estuary saltmarsh plant and sea-grass communities 169
Stephen Waite
9.1 Introduction 169
9.2 Saltmarsh development 170
9.3 Intertidal seagrass communities 172
9.4 Saltmarsh vascular plants and communities 175
9.5 The origin and spread of *Spartina* 177
9.6 Regional saltmarsh communities 179
9.7 Saltmarsh vegetation of the greater Thames 186

10 Estuaries: towards the next century 191
Alasdair McIntyre
10.1 Introduction 191
10.2 Early work on estuaries 193
10.3 Concerns for the future 195

Appendices 199
A Summary of data for selected sites along the Thames estuary 199
B Thames estuary species list 200

References 235
Index 245

Contributors

Dr Martin J. Attrill
Department of Biological Sciences
University of Plymouth
Drake Circus
Plymouth PL4 8AA
UK

Dr R.G. Bailey
Division of Life Sciences
King's College
University of London
Campden Hill Road
London W8 7AH
UK

Dr A.W. Bark
Division of Life Sciences
King's College
University of London
Campden Hill Road
London W8 7AH
UK

Dr Chris Gordon
Volta Basin Research Project
University of Ghana
Legon
Ghana

Professor Jim Green
17 King Edwards Grove
Teddington, Middlesex TW11 9LY
UK

Dr David John
Botany Department
The Natural History Museum
Cromwell Road
London SW7 5BD
UK

Jane Kinniburgh
Environment Agency
Rivers House
Waterside Drive
Aztec West
Almondsbury
Bristol BS12 4UD
UK

Dr Steve Lee
Division of Life Sciences
King's College
University of London
Campden Hill Road
London W8 7AH
UK

Professor Alasdair McIntyre
63 Hamilton Place
Aberdeen AB15 5BW
UK

Dr Margaret Munro
Division of Life Sciences
King's College
University of London
Campden Hill Road
London W8 7AH
UK

R. Myles Thomas
Environment Agency
Apollo Court
2 Bishops Square Business Park
St Albans Road West
Hatfield, Hertfordshire AL10 9EX
UK

Dr Derek Tinsley
National Centre for Ecotoxicology and Hazardous Substances
Environment Agency
Evenlode House
Howbery Park
Wallingford
Oxfordshire
OX10 8BD
UK

Ian Tittley
Botany Department
The Natural History Museum
Cromwell Road
London SW7 5BD
UK

Dr Stephen White
Biology Section
University of Brighton
Cockcroft Building
Moulsecoombe
Brighton BN2 4GJ
UK

Professor Phil Whitfield
Head: Division of Life Sciences
King's College
University of London
Campden Hill Road
London W8 7AH
UK

Acknowledgements

Chapter 2

D.T. would like to express his gratitude to Jane Kinniburgh and Peter Lloyd for helpful discussions and their constructive comments on an early draft of this chapter. Any errors that remain are entirely due to the author. The views expressed in this chapter are those of the author and not necessarily those of the Environment Agency.

Chapter 3

J.K. wishes to acknowledge the help of Mr J.F. Payton, Dr J. Bennett and Mr M. Tinsley in the preparation of the figures and Miss D. McCallum in typing the text. This chapter is published with the permission of Mr L.D. Jones, Regional General Manager of the National Rivers Authority, Thames Region (now Environment Agency). The views expressed are those of the author and do not necessarily reflect those of the Environment Agency.

Chapter 4

I.T. and D.J. wish to thank J.H. Price for field information, Dr Chris Maggs for comments on the taxonomy of *Polysiphonia urceolata / P. subtilissima*, and Gordon Paterson for help with numerical analyses.

Chapter 5

The work presented here was supported by a grant to C.G. from the Association of Commonwealth Universities. The National River Authority (Thames Region) was most helpful in providing information from their own sampling programmes; much of this assistance has been a result of the efforts of Dr D. Tinsley. Professor J. Green of Queen Mary College, University of London, and Dr G. Boxshall of the Natural History Museum assisted in the identification of zooplankton. Valuble comments were made on the manuscript by Dr Roland Emson, Professor Brian Gardiner, and Dr Peter Williams.

CHAPTER 1

Introduction

Jim Green

1.1 THE THAMES ESTUARY: A PERSONAL VIEW

For almost 50 years I have lived close to the Thames estuary (Figure 1.1). In the early 1950s, from our flat in Pimlico, we could walk along the embankment opposite Battersea Power Station. At low tide, the exposed mud had large red patches caused by the haemoglobin in innumerable tubificid worms. These formed the basis of a minor trade. Men with waders and sieves would collect the worms and sell them to the aquarium trade as food for fishes. The superabundance of these worms depended on the gross organic pollution of this reach of the Thames.

Towards the end of the 1950s we moved to Teddington, within a few minutes' walk from the lock. Casual observation of the birds on the river indicates an increase in the piscivores over the last 20 years. Cormorants (*Phalacrocorax carbo*) are regularly seen, and grey herons (*Ardea cinerea*) are more abundant. On a good day it is possible to see up to eight herons around the weir and below the lock. Another bird that is now commonplace in the area is the great-crested grebe (*Podiceps cristatus*). The regular mallard (*Anas platyrhynchos*) and Canada geese (*Branta canadensis*) are sometimes joined by tufted duck (*Aytha fuligula*) and mandarin (*Aix galericulata*).

In our early days at Teddington there were several tall Lombardy poplars near the lock. These served as markers for swarms of emerging chironomids. In the early summer, these flies were so abundant that the tops of the trees appeared to be emitting smoke as the swarms wafted about in the breeze. These swarms are not so frequent or so large now, probably due to a combination of the clean up of the river and the increase in fish which prey upon the larvae.

One spectacular invertebrate that has increased immensely during my time on the Thames is the Chinese mitten crab, *Eriocheir sinensis*. This, as its name implies, is a native of Chinese rivers, but has been accidentally introduced into Europe. It was first recorded from the Thames in 1935, but

A Rehabilitated Estuarine Ecosystem. Edited by Martin J. Attrill.
Published in 1998 by Kluwer Academic Publishers, London. ISBN 0 412 49680 1.

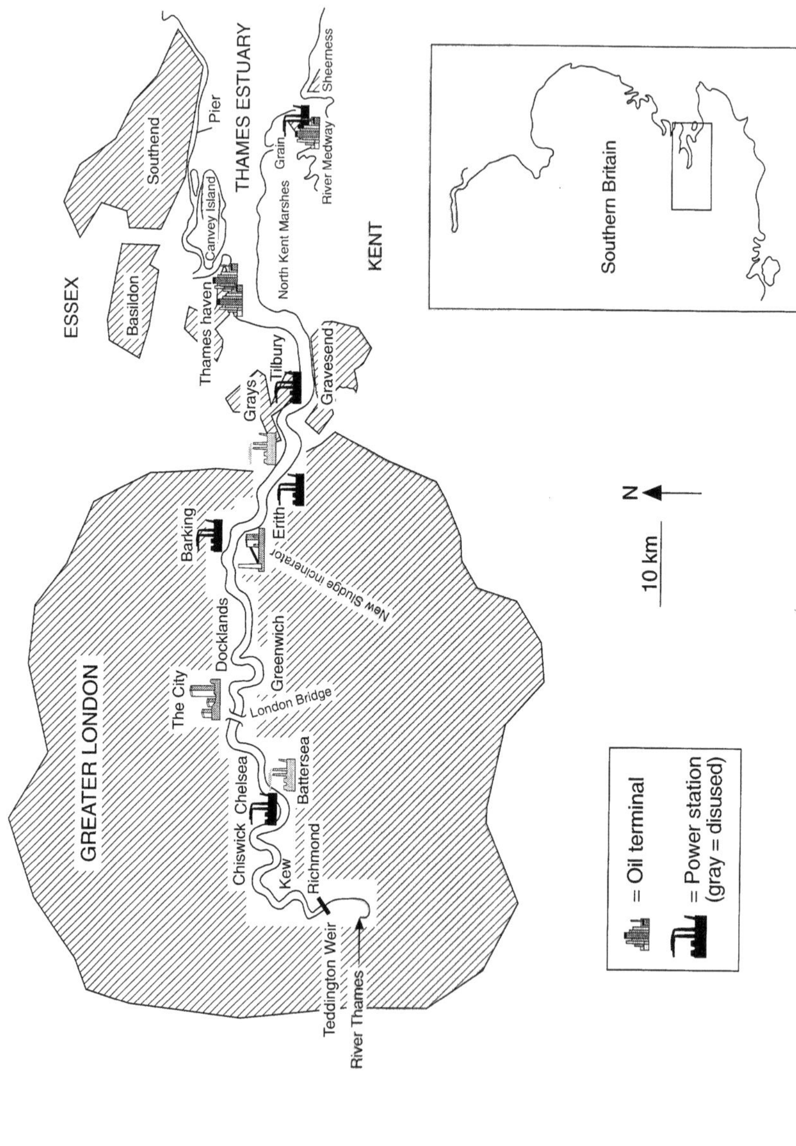

Figure 1.1 The Thames estuary region, with surrounding urban areas.

for many years there were no further records. I saw my first specimen in Hampton Court Park in 1961. Since then there have been increasing numbers of records, and it is now found abundantly on the filter screens at the cooling water intakes of some of the power stations on the estuary (Attrill and Thomas, 1996). It has also been found in Bushy Park, having travelled up a minor tributary of the Thames.

1.2 CLEANING UP THE THAMES AND THE ROLE OF THIS BOOK

That these obvious changes have taken place during a single lifetime is an indication of major alterations in the conditions within the estuary. Details of these changes require regular sampling and critical analysis. A good account of the birds and their food has been given by Harrison and Grant (1976) and Alwyne Wheeler (1979) has documented the return of fish to the tidal Thames. Both these books were particularly concerned with the initial 'Great Clean Up' which was brought about mainly by improvements in sewage treatment.

There are many different ways of looking at an estuary. For instance, in 1936 H. Muir Evans, the Vice-Commodore of the Royal Norfolk and Suffolk Yacht Club, published *A Short History of the Thames Estuary*. It was concerned almost entirely with the distribution and naming of sand banks – a matter of importance if you are navigating in the area.

The present book seeks to bring together certain aspects of the situation in the Thames estuary, and to update the changes that have occurred. It is not possible to cover every aspect, but the various chapters cover a wide range, from physicochemical factors to algae and fish. Of necessity, each field requires its experts, but eventually, for a proper understanding of an estuary there must be an integration of the varied data.

In October 1971, a meeting was held at the Zoological Society of London to inaugurate a new scientific body, provisionally entitled the Estuarine and Brackish-water Biological Association. In a brief but wise summarizing review, the late Don Arthur argued strongly that there was a need for hydrographers, biologists, sedimentologists, chemists, bacteriologists, engineers and members of other disciplines to integrate their research. He proposed the deletion of the word 'Biological' from the name of the Association, to encourage other scientists to join. The name was promptly changed, and the Estuarine and Brackish-water Sciences Association has since played a leading role in the integration of our knowledge of these complex systems. This in no way diminishes the importance of biological studies, but requires that they take due cognizance of the physical and chemical aspects of the environment. A perusal of the various chapters in this book will show that biologists have received the message, but Professor McIntyre's final chapter indicates new messages. If our estuaries are to be kept in a healthy condition there must be cooperation beyond the scientists. The dramatic changes in the Thames show that rehabilitation can take place if the political will is there.

CHAPTER 2

The Thames estuary: a history of the impact of humans on the environment and a description of the current approach to environmental management

Derek Tinsley

2.1 INTRODUCTION

The environment and ecology of the Thames estuary are closely linked with the activities of humans. The estuary was initially used by humans for food, then for transport and communication and more recently as a location for industry and development. With the growth of London in the nineteenth and twentieth centuries, parts of the estuary became polluted and devoid of much aquatic life. In addition, valuable estuarine habitat was lost to farming, residential and industrial development. Extensive remedial action has resulted in an improved water quality and much wildlife has returned. The estuary can again be considered as a valuable natural resource. However, the threat to the environment from human activity still remains and the importance of careful environmental regulation is clear for all to see.

The aim of this chapter is to provide an introduction to the Thames estuary, its environmental history and current environmental management. After a basic description of the estuary as it is today, a brief insight into the estuarine environment, and the presence and diversity of estuarine

A Rehabilitated Estuarine Ecosystem. Edited by Martin J. Attrill.
Published in 1998 by Kluwer Academic Publishers, London. ISBN 0 412 49680 1.

organisms before the growth of London after 1800, will be given. This will be followed by a short history of the impact of humans on the environment between 1800 and the present day, covering both water quality and habitat loss within the estuary. Finally, a brief description of the current approach to environmental management will be provided.

2.2 DEFINITION OF THE STUDY AREA

The question, 'Where does the Thames estuary begin and end?' is not an easy one to answer. A number of upstream and downstream limits can easily be fixed, but few of these will have any biological significance. For the purposes of this book, the term 'the Thames estuary' will be used to describe an area of tidal water which stretches from Teddington Weir seaward to a line just to the east of Shoeburyness (Figure 2.1).

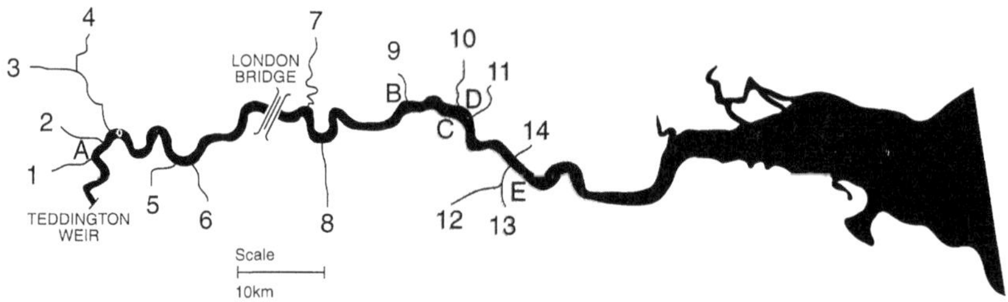

Figure 2.1 A map of the Thames estuary showing the location of tributaries and the major sewage treatment works. **Tributaries**: **1**. River Crane; **2**. Duke of Northumberland's River; **3**. Grand Union Canal; **4**. River Brent; **5**. Beverley Brook; **6**. River Wandle; **7**. River Lee; **8**. River Ravensbourne; **9**. River Roding; **10**. River Rom/Beam; **11**. River Ingrebourne; **12**. River Cray; **13**. River Darent; **14**. Mar Dyke. **Major sewage treatment works**: **A**. Mogden; **B**. Beckton; **C**. Crossness; **D**. Riverside; **E**. Long Reach.

Teddington Weir was first built in 1811 to ensure that a minimum depth of water was maintained in the river Thames upstream for navigational purposes. The weir is now the upstream tidal limit of water in the estuary. The seaward line referred to runs from Haven Point on the north shore in Essex to Warden Point on the Isle of Sheppey in Kent and was the seaward boundary used by the Port of London Authority and the seaward pollution control limit used by the Environment Agency (EA), Thames Region (previously the National Rivers Authority).

The term 'the **greater** Thames estuary' will be used in the account of habitat loss and in the description of the current approach to habitat management in the estuary. It is used to describe an area which encompasses the discharges of five large rivers: the Thames, the Medway, the Crouch, the Blackwater and the Colne.

2.3 BASIC DESCRIPTION OF THE THAMES ESTUARY

The Thames estuary is an example of what is termed a 'coastal plain estuary' (Pritchard, 1955), formed as a result of a rise in the level of what we now call the North Sea, flooding the lower Thames Valley. It has a V-shaped (sometimes referred to as a trumpet-shaped or bell-shaped) outline characteristic of many coastal plain estuaries.

2.3.1 Width and depth

The present width of the estuary ranges from approximately 90 m at its upstream limit, Teddington, to approximately 240 m at London Bridge. The latter is commonly used as a reference point (a datum zero) for distance measurements along the estuary and so, for example, Teddington Weir is frequently described as being 31 km upstream of London Bridge. As you move downstream, the estuary widens further. At the new Dartford Bridge, some 33 km below London Bridge (Figure 1.1), the estuary is approximately 800 m in width and at the popular north shore seaside town of Southend-on-Sea, its width has increased again to some 7 km.

The depth of water in the estuary varies with the state of the tide. However, both the depth and the tidal movement of water between Teddington and Richmond are in part regulated by a half-tide barrier situated just downstream of Richmond Bridge and some 25 km upstream of London Bridge. This barrier was constructed in 1828, as with Teddington Weir, to ensure a minimum depth of water upstream for navigational purposes. As the tide ebbs, the barrier gates are lowered into position to maintain the water upstream at about the half-tide level (1.7 m above Ordnance datum Newlyn). On the flood tide, the barrier gates are raised again when the water level downstream has reached that of the impounded water. Further details of the operation of the Richmond half-tide barrier are given in Bowen and Pinless (1977). The barrier is not operated for a few weeks each year to allow the tidal flow to remove silt which tends to settle from the water in the less turbulent conditions upstream of the barrier.

Some 15 km downstream of London Bridge, situated at Woolwich, is the only other barrier across the main Thames estuary: the better known – and far more extensive – Thames Barrier. This was completed in 1972 as an integral part of London's flood defences. Fortunately, the threat to London posed by tidal surges is infrequent and so for most of the time the Thames Barrier remains open, not affecting the movement of water, silt and aquatic organisms within the estuary. Further information on the Thames Barrier can be found in Gilbert and Horner (1992). A number of other flood defence barriers exist across tidal tributaries and creeks within the estuary. The location of these will be described later in this chapter.

2.3.2 Freshwater inputs

The estuary upstream of Southend receives freshwater inputs from the main river Thames and 14 tributaries. The names and position of these tributaries are given in Figure 2.1 and the volumes of freshwater discharged from them are given in Table 2.1. The estuary also receives freshwater inputs from a number of sewage treatment works. The positions at Beckton, Crossness, Long Reach, Mogden and Riverside sewage treatment works are also given on Figure 2.1.

The data in Table 2.1. suggest that the freshwater flow entering the tidal Thames from the main river above Teddington greatly exceeds that of all the freshwater tributaries added together. However, during the summer, the flow over the weir can drop to below 10 cumecs. Under these conditions, the sewage effluents from Beckton and Crossness sewage treatment works become the largest freshwater inputs to the estuary.

2.3.3 Current use of the estuary

The Thames estuary today is used by humans for a variety of purposes. Considerable volumes of water are abstracted from the estuary for use in industrial cooling, although much of this water is eventually returned near to the point of abstraction. In particular, there are a number of large power

Table 2.1 Freshwater inputs to the tidal Thames – main river and tributaries

Source	*Mean Flow: 1978–1988 (Cumecs)*
Main river Thames	67.8
London rivers	5.2
Crane	
Duke of Northumberland's	
Brent	
Beverley Brook	
Wandle	
Ravensbourne	
River Lee	6.9
Essex rivers	2.9
Roding	
Beam	
Ingrebourne	
Mar Dyke	
River Darent	1.1

stations sited on both the north and south shores requiring cooling water. Industry also uses the estuary as a route for disposal of liquid effluents. As indicated in section 2.3.2, the largest volumes of effluent entering the estuary are from sewage treatment works. These effluents contain both domestic and industrial waste and, as will be discussed later in this chapter, they have a significant influence on water quality.

Despite the world decline in shipping, London is still an important port. A dredged channel allows ships access to Tilbury Docks and various upriver berths from the North Sea. Commercial fishing is limited these days within the estuary. However, there is an important cockle fishery to the east of Southend, and some sprat, herring, sole and eel are also caught for commercial gain. Finally, the estuary has a public amenity value. There are three EC designated bathing beaches in the Southend area, and sailing, windsurfing and angling are all popular.

2.3.4 Estuary habitats

A number of different habitats are well represented within the estuary. There are extensive subtidal and intertidal areas with gravel, mud and sandy substrates, particularly around the Southend area, where the tide can retreat up to 4 km. Saltmarsh and coastal grazing marsh are limited in towards London, but are much more extensive to the east of Southend. A saline lagoon habitat exists on the south shore at Cliffe near to Gravesend. Sand dune and rocky shore habitats are poorly represented, although the numerous piers and extensive flood defence walls in towards London provide some hard substrates for the attachment of aquatic plants and animals.

2.4 THE THAMES ESTUARY BEFORE 1800

The extensive growth of London in the nineteenth and twentieth centuries resulted in a drastic decline in the environmental quality of parts of the Thames estuary. Before describing this decline and the remedial action taken to improve water quality in the estuary, it may be of value to consider what the Thames estuary was like before 1800. Wood (1982) attempted such an exercise in his book, *The Restoration of the Tidal Thames*, but he concentrated more on the social history of the estuary than on its natural history. It is hoped that the present account will provide a baseline against which the changes in water quality and loss of habitat described later in this chapter can be compared.

2.4.1 Methodology

Amongst the sources of information that can be used to provide the insight described above, published literature and written records from the past are probably the most useful. However, searching out and reviewing historical documents is time consuming and a highly skilled job. Fortunately, some historical accounts reviewing past literature and records of relevance already exist. An account of tributaries which once joined the tidal Thames, the so-called Lost Rivers of London (Barton, 1962), and accounts of fishes and fisheries of the tidal Thames before the growth of London in the nineteenth and twentieth centuries (Wheeler, 1958, 1979) are of particular value in this respect.

2.4.2 The estuary environment before 1800

The river Thames was thought to be tidal as far upriver as Staines before the construction of Teddington Weir and the Richmond half-tide barrier in the early part of the nineteenth century (Herbert, 1966). The estuary between Teddington and Richmond would have been shallower than it is today with more extensive gravel banks exposed to the air at low water and adjacent water meadows flooded at times of high freshwater flow.

In towards central London, the estuary would have been wider before the construction of embankments confining the tidal channel. Furthermore, there would have been great expanses of marshland, culminating in true saltmarsh as one moved towards the mouth of the estuary.

Many more freshwater tributaries joined the tidal Thames in the past. Barton (1962) lists 14 such tributaries in his account, *Lost Rivers of London*. The largest of these was the Fleet River, with twin sources at Hampstead Heath and Highgate to the north of London and its confluence with the tidal Thames just to the west of the City of London. Such tributaries would have provided valuable habitats for wildlife.

Finally, water quality in the main estuary would have generally been of a high standard. Unfortunately, there is no chemical data to support this view. It was not until the voyage of the *Challenger*, from 1873 to 1876, that marine chemistry became a science (Riley and Chester, 1971) and not until 1882 that the first water quality surveys were carried out for the Thames estuary (Water Pollution Research Laboratory, 1964). However, the presence of a good diversity of fish species, including salmonids, before 1800 (Wheeler, 1958, 1979) would tend to suggest that the concentrations of dissolved oxygen were generally high.

2.4.3 The presence and abundance of estuarine organisms before 1800

Reviews of the commercial fisheries which existed in the estuary before 1800 provide information on the diversity and abundance of aquatic life. Wheeler (1958, 1969, 1979), describes commercial fisheries for fin-fish such as lampern (*Lampetra fluviatus*), twaite shad (*Alosa fallax*), smelt (*Osmerus eperlanus*), flounder (*Platichthys flesus*), eel (*Anguilla anguilla*) and, in the more downstream reaches, sprat (*Sprattus sprattus*) and herring (*Clupea *harengus*). The financial value of these fisheries and records of the number of fish caught suggest that the species listed were present in abundance. Other species of less commercial interest, such as salmon (*Salmo salar*), sea trout (*Salmo trutta*), lamprey (*Petromyzon marinus*) and sturgeon (*Acipenser sturio*), were reported as being present in the estuary but not in such large numbers. Records of the freshwater fish population of the tidal Thames prior to 1800 are few and far between.

The existence of commercial fisheries for brown shrimp (*Crangon crangon*), pink shrimp (*Pandalus montagui*), oyster (*Ostrea edulis*) and starfish (*Asterias rubens*) towards the mouth of the estuary suggests that these invertebrates were also present in considerable numbers. The starfish caught were not used for human consumption, but sold to local farmers for use as manure (Wheeler, 1979).

2.5 A HISTORY OF THE IMPACT OF HUMANS ON THE ENVIRONMENT (1800 TO THE PRESENT DAY)

2.5.1 Water quality

Since humans first settled by the tidal Thames back in Roman times, there has probably always been some pollution of the estuary and its tributaries. However, it was not until the nineteenth century that this pollution became so bad as to cause a major decline in the water quality of the main estuary. The pollution history of the estuary during the nineteenth and twentieth centuries has been well documented by a number of authors (Potter, 1971; Gameson and Wheeler, 1977; Wheeler, 1979; Wood, 1980, 1982; Andrews and Rickard, 1980; Andrews, 1984). In reading these accounts, it becomes clear that since 1800 there have been two periods of very poor water quality, during which stretches of the estuary were completely devoid of oxygen, with each period being followed by an improvement in water quality as a result of extensive remedial action.

2.5.2 The decline in water quality in the nineteenth century

The first major decline in estuarine water quality took place during the first half of the nineteenth century with the growing importance of London as a centre of commerce and the growth of new industries. The population of London grew from just over one million people at the beginning of the century to 2.75 million people by about 1850 (Wood, 1982). With the invention of the water closet in 1810, a high proportion of the human waste from London found its way into the estuary via tributaries. This waste was readily degraded by microorganisms using up the available oxygen in the water. At the same time, industrial pollution was contributing to the decline in water quality. In particular, the waste liquors from the production of coal-gas were being discharged directly into the estuary. As a result, much of the estuary in towards London was devoid of life and notorious for the smell of hydrogen sulphide. In the summer of 1858, water quality was so bad that the windows of the Houses of Parliament had to be draped with curtains soaked in chloride of lime so that members could breathe: 1858 became known as 'The Year of the Great Stink'.

2.5.3 Interceptor sewers for London and chemical treatment of the sewage

In an attempt to improve the management of London's drainage, new legislation was drawn up in 1855 and a Metropolitan Board of Works was created to deal with the problems. The Board's Chief Engineer, Sir Joseph Bazalgette, was given the task of designing a sewerage system that would prevent sewage from entering the estuary in or near the metropolis. He proposed building interceptor sewers on the north and south shores of the estuary. These would carry sewage east of London to Beckton on the north shore and to Crossness on the south shore. There were no plans to treat the sewage at these locations, but just to store it in reservoirs ready to be discharged via two large outfalls into a 3 km stretch of Barking Reach on each ebb tide. After some debate over the proposals, work on constructing the interceptor sewers was eventually started in 1858 and completed in 1864.

The construction of interceptor sewers helped to reduce the pollution problem in towards central London, but the discharge of the untreated sewage into Barking Reach resulted in an immediate decline in water quality in this part of the estuary. Local complaints led to the setting up of a Royal Commission on Metropolitan Sewage Discharge in 1882, charged with looking into the problems. The Royal Commission recommended chemical precipitation of the sewage during sedimentation to reduce its biological oxygen demand. The resultant liquid could then be discharged to the estuary on the ebb tide and the watery sewage solids dumped in the outer estuary by boat. Various methods of chemical precipitation were tried in a 31-year period between 1884 and 1915.

The Metropolitan Board of Works was superseded by London County Council in 1889. The need for more interceptor sewers was recognized to accommodate the continuing growth of London and to provide a greater capacity in the sewerage system to cope with storm water.

The combined effects of chemical precipitation of the sewage reducing its biological oxygen demand and the additional interceptor sewers reducing the frequency of storm discharges of sewage into the estuary in central London resulted in improvements in water quality within the estuary. By the turn of the century fish had returned to reaches which had previously been devoid of much aquatic life.

2.5.4 The decline in water quality in the twentieth century

The improvements in water quality achieved by the end of the nineteenth century lasted until the First World War (1914–1918). Thereafter, water quality in the estuary started to decline again. Wood (1982) has identified three distinct phases in water quality in the Thames estuary during the first half of the twentieth century.

The first of these was one of relatively good water quality. Chemical analysis of water samples taken between 1893 and 1910 showed that the concentrations of dissolved oxygen remained above 25% saturation in all places, despite a marked increase in the population of London during this period.

In the second phase, between 1915 and 1930, the concentrations of dissolved oxygen fell almost to zero within a stretch of the estuary 10 km upstream and 10 km downstream of the north and south outfalls. The continuing increase in the population of London, particularly after the First World War, and the decision to discontinue chemical treatment of the sewage at the main outfalls are thought to be the likely causes of this marked decline.

During the third phase, between 1935 and 1950, a zone of oxygen depletion covering a greater distance along the estuary than that observed in the second phase was found to exist. Wood attributed this to the spread of London's population and the discharge of effluents into the estuary and its tributaries from a number of small sewage treatment works built in the new suburbs. Some attempts were made to reduce the number of these small regional works and to plan for the increasing population. In 1935, Mogden sewage treatment works was constructed to replace 28 small works. However, further progress towards fewer, larger works providing better standards of treatment was stopped by the beginning of the Second World War (1939–1945).

The pollution problems during this third phase were made worse by the increased use of non-biodegradable ('hard') synthetic detergents after the Second World War. These detergents produced persistent foams which significantly reduced the rate of oxygen exchange at the water surface. The detergents also reduced the efficiency of the few secondary treatment plants operating at that time. The 'hard' surfactants in the detergents inhibited the microorganisms involved in the secondary treatment process. One of the first such plants had been installed at the North Outfall Works in 1931. The non-biodegradable detergents were later to be replaced by more readily biodegradable substitutes, but not before they had significantly contributed to the decline in estuarine water quality.

The warm water discharged into the estuary from power stations also added to the pollution problems during the third phase. The increased water temperature presented a direct threat to the existence of cold-water species and in addition it reduced the oxygen-carrying capacity of the water.

By the end of the war, the estuary was anaerobic all year round near the main outfalls. In the dry summer of 1949, records show that an anaerobic zone stretched for some 42 km.

2.5.5 The Pippard Report and the Water Pollution Research Laboratory study

By the middle of the twentieth century, there was again great public concern about water quality in the Thames estuary. In 1951, the government of the day set up a committee, under the chairmanship of Professor Pippard, to look into the effects of the various discharges on water quality in the estuary.

The Pippard Committee recognized that a detailed scientific survey of water quality was required before recommendations could be made on the remedial action necessary to bring about improvements. Such a survey was carried out by the Water Pollution Research Laboratory under the direction of the Thames Survey Committee. The latter had been set up a few years earlier to investigate the causes of silting in the estuary near Barking.

The final report produced by the Water Pollution Research Laboratory provided a very detailed description of estuary water quality (Water Pollution Research Laboratory, 1964). It also identified the various sources of pollution within the estuary and provided an estimate of the contribution each made towards the overall polluting load.

The Pippard Committee recommended that water quality should be improved to an extent whereby the water was no longer a nuisance to the public (Ministry of Housing and Local Government, 1961). The nuisance referred to was chiefly the smell of hydrogen sulphide. The study carried out by the Water Pollution Research Laboratory showed that hydrogen sulphide was not produced if there was always a detectable quantity of dissolved oxygen or nitrate in the water. The committee also considered the need to improve water quality even further so as to allow the passage of migratory salmonids, but concluded that the costs involved would greatly exceed the benefit.

Having identified the target for water quality in the estuary, the committee then went on to recommend actions that would need to be taken if the target was to be achieved. These included carrying out improvements to the major sewage treatment works discharging into the estuary. In fact, before the Pippard Committee Report was completed, work to extend Mogden and Beckton Sewage Treatment Works had begun. In 1962, Deephams Sewage Treatment Works was built on the lower part of the River Lee catchment, replacing 14 smaller works. This led to an improvement in quality of the freshwater discharge from the Lower Lee entering the estuary. In addition, a full treatment plant was completed at the Crossness Works in 1963. Details of the type of sewage treatment plant installed at each of these works can be found in Wood (1982). These improvements resulted in a reduction in the overall pollution load entering the estuary and by 1966, aerobic conditions were achieved all year round in the tidal Thames.

The Pippard Committee also recommended that the Port of London Authority should take on a more encompassing role as the pollution control authority for the Thames estuary. The Authority had held statutory responsibilities for the conservancy of the tidal Thames since it was set up in 1909. The Port of London (Consolidation) Act, 1920, charged the authority with maintaining the flow and purity of the water in the estuary, but it was the London County Council that actually carried out regular water quality surveys. Furthermore, the system of setting and policing limits placed on discharges to the estuary was weak. In fact, under the 1920 Act, the London County Council was exempt from the control of the Port of London Authority with respect to pollution control matters.

The job of monitoring water quality within the estuary was transferred to the Port of London Authority. The Port of London Act, 1964, gave the authority statutory pollution control powers more akin to those previously exercised by the various river authorities under the Rivers (Prevention of Pollution) Acts, 1951 and 1961. In 1965, the London County Council was abolished and replaced by the Greater London Council. The responsibility for London's sewerage system and the operation of its major sewage treatment works such as Beckton and Crossness was automatically transferred to the new council.

2.5.6 The Royal Commission on Environmental Pollution (Third Report)

In the early 1970s, the Royal Commission on Environmental Pollution turned its attention to pollution in British estuaries and coastal waters (Royal Commission on Environmental Pollution, 1972). It recommended that toxic and non-biodegradable substances should be excluded from discharges. It also recommended that the government should adopt two biological criteria for the management of estuarine waters:

- the ability to support on the mud bottom the fauna essential for sustaining sea fisheries;
- the ability to allow the passage of migratory fish at all states of the tide.

A third recommendation was that water quality in estuaries should be maintained by setting up 'pollution budgets', these being the maximum pollution load that an estuary can accept and still meet the biological criteria listed above.

2.6 WATER QUALITY MANAGEMENT DURING THE THAMES WATER AUTHORITY ERA

The Control of Pollution Act, 1974, brought the jobs of regulating water quality and managing London's major sewage treatment works together to be the responsibility of one new organization called the Thames Water Authority. As a result, the task of interpreting the Royal Commission's recommendations fell largely to this newly formed organization.

2.6.1 Water quality objectives

The Thames Water Authority set about the task of integrating the recommendations of the Royal Commission into a plan for managing water quality in the Thames estuary. The estuary was initially divided into four reaches for the purposes of water quality management. A set of water quality objectives and standards for the concentration of dissolved oxygen were formulated for each reach (Table 2.2).

The standards were based on concentrations of dissolved oxygen in the water as the discharge of biodegradable organic matter and oxidizable ammonia into the estuary, primarily from sewage treatment works, was considered to be the most important source of pollution. It was intended to combine the two freshwater reaches once water quality in the Barnes to London Bridge reach had improved following the upgrading of Long Reach Sewage Treatment Works.

The best water quality in the estuary was to be found in the freshwater and marine reaches. The dissolved oxygen standard for the freshwater reach between Teddington and Barnes was a minimum concentration of 40% saturation for 95% of the time (also expressed as a quarterly average of > 60% saturation). This is equivalent to that of a National Water Council Class 2 river (National Water Council, 1976). In the absence of guidance on suitable dissolved oxygen standards for the marine reach, the value for a National Water Council Class 1B freshwater river was used (Andrews *et al.*, 1983), i.e. a minimum concentration of 60% saturation for 95% of the time (also expressed as a quarterly average of > 80% saturation).

The poorest water quality was to be found in the euryhaline reach which received the effluents from a number of sewage treatment works, including Beckton and Crossness. Two standards were set for this reach. The first of these was a minimum concentration of dissolved oxygen of 10% saturation for 95% of the time. In effect, this was the standard recommended by the Pippard Committee, i.e. that the estuary should be aerobic all year round and therefore no longer a nuisance to the public. The second standard was a concentration of dissolved oxygen of > 30% saturation expressed as a quarterly mean. This was an attempt to demonstrate that a water quality adequate for the passage of migratory fish could be achieved at all states of tide, as recommended by the Royal Commission. The figure

Table 2.2 Thames Water – estuary water quality objectives and standards

Reach	*Water quality objective*	*Standard*
Teddington to Barnes	• Passable to migratory fish • Maintenance of coarse fishery • Water available for potable supply after advanced treatment	NWC Class 2 • DO > 40% saturation for 95% of the time (quarterly average DO of > 60% saturation)
Barnes to London Bridge	• Absence of nuisance through anaerobicity • Passable to migratory fish	• DO > 10% saturation for 95% of the time (quarterly average DO of > 30% saturation)
London Bridge to Canvey Island	• Absence of nuisance through anaerobicity • Passable to migratory fish	• DO > 10% saturation for 95% of the time (quarterly average DO of > 30% saturation)
Canvey Island to Seaward Limit	• Quality should be suitable for the whole life cycle of marine organisms	• DO > 60% saturation for 95% of the time (quarterly average DO of > 80% saturation)

DO = dissolved oxygen

of 30% saturation was based on laboratory data for salmon quoted in the Pippard Report. However, there was some debate as to whether or not this standard should be expressed as a minimum of 30% saturation instead of as a quarterly mean (Andrews *et al.*, 1983).

2.6.2 A pollution budget for dissolved oxygen

The results of surveys of the concentrations of dissolved oxygen at different points along the estuary throughout the year revealed an oxygen minimum during the summer which occurred some 28 km below London Bridge. It was suggested that by controlling the concentration of dissolved oxygen at this oxygen minimum (known as the critical point), compliance with the various water quality standards for dissolved oxygen throughout the estuary could be achieved (Lloyd and Cockburn, 1983). Therefore, a pollution budget was constructed for the estuary in terms of the concentration of dissolved oxygen at this critical point (Figure 2.2). A method of dividing the available budget between the different polluting inputs was devised and a mathematical model was used to formulate limits (known as discharge consents) on the various point source discharges likely to effect the concentration of dissolved oxygen at the critical point (Lloyd and Cockburn, 1983).

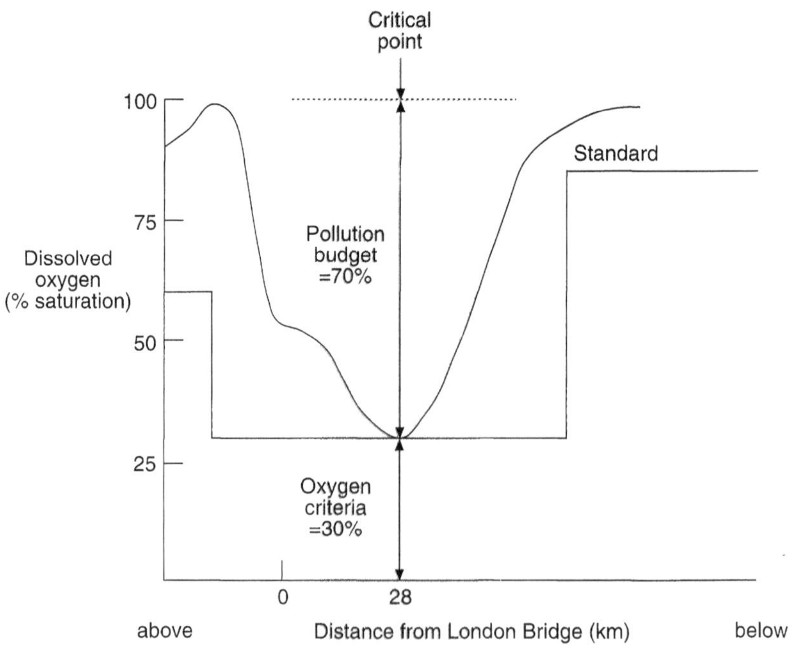

Figure 2.2 Thames estuary pollution budget for dissolved oxygen.

2.6.3 Improved water quality resulting from better sewage treatment

Further improvements to Beckton Sewage Treatment Works were completed in 1974 and modifications to the Riverside works, which enabled it to deal with trade effluents containing inhibitors of nitrification, were completed about the same time. Finally, major improvements to Long Reach Sewage Treatment Works were completed in 1979. Again, details of the type of sewage treatment plant installed at each of these works can be found in Wood (1982).

As in the 1960s, this major expenditure on sewage treatment works resulted in improvements in water quality within the estuary. Andrews and Rickard (1980) reported a greater diversity and increased abundance of fish and macroinvertebrate species during what they describe as a second phase in the rehabilitation of the Thames estuary ecosystem.

In the 1980s, arrangements were made for a temporary improvement in the quality of the final effluents from sewage treatment works such as Mogden and Beckton each summer, in order to improve compliance with the dissolved oxygen standards described earlier.

2.6.4 The threat to water quality from storm discharges

In addition to controlling the pollution load entering the estuary from the major sewage treatment works and industry, there was also a need to combat a threat to water quality from storm discharges. The latter term describes the cumulative effect of surface runoff, increased flows from sewage treatment works and storm sewage discharges from the combined drainage of inner London which occur at times of heavy localized rainfall. The storm discharge problem was not a new one, but its importance had been masked by the overall poor water quality in the estuary before the improvements to the main sewage treatment works. Although relatively short in duration, storm discharges were highly polluting, reducing the concentrations of dissolved oxygen in the estuary to a level that was lethal to aquatic life. In 1973 and 1977, storm discharges resulted in major fish kills in the estuary (Lloyd and Whiteland, 1990).

Solutions to the storm discharge problem were put forward and 'in-river oxygenation' was considered to be the most cost-effective option. In 1979, a prototype oxygenation unit constructed on a barge was purchased. The prototype demonstrated that in-river oxygenation could be used with some success to tackle the storm discharge problem and 10 years later a second purpose-built unit with a greater capacity for oxygen addition (the 'Thames Bubbler') was brought into operation (Lloyd and Whiteland, 1990). A description of both units and their use in connection with a network of automatic water quality monitoring stations situated at intervals along the estuary is described by Griffiths and Lloyd (1985) and Lloyd and Whiteland (1990).

Another measure used to combat the storm discharge problem was to increase the flow of well oxygenated freshwater over the weir at Teddington by suspending abstraction for potable supply upstream. However, there was a time delay between taking this action and water quality in the estuary benefiting from the measure described.

2.6.5 Control of toxic, non-biodegradable substances

Although oxygen depletion was considered to be the most important pollution problem in the estuary, the discharge of toxic, non-biodegradable substances was not neglected. In line with the recommendations of the Royal Commission, more stringent trade effluent controls were imposed on industry by the Thames Water Authority. Andrews (1984) reported reductions in the concentrations of various organohalogen pesticides, PCBs, metals and detergents in samples of water and fish taken from the estuary as a result of these measures.

The introduction of the EC Dangerous Substances in Water Directive (76/464/EEC) in 1976 resulted in further pressure to control the discharge of toxic and non-biodegradable substances into the estuary. The directive provided two lists of dangerous substances (List 1, the 'Black List', and List 2, the 'Grey List') selected on their toxicity, persistence and bioaccumulation. The Thames Water Authority was responsible for consenting to the discharge of these substances into the estuary and for monitoring compliance with receiving-water environmental quality standards, in addition to regulating the quality of trade effluents entering its sewage works.

2.7 HABITAT LOSS AS A RESULT OF FLOOD DEFENCE WORK AND LAND-CLAIM

Over the years there has been a continued loss of habitat within the Thames estuary. In particular, the once extensive marshland on both the north and south shores has been reclaimed behind embankments. Most of the early reclaimed land was used as grazing marsh. Much of this grazing marsh was later converted for use as arable land or for industrial and residential development. Ekins (1990) reported a 64% loss of grazing marsh in the greater Thames estuary between 1930 and 1980. Conversion to arable agriculture was the main cause, although industrial and residential development has also had a significant impact.

The most extensive loss of estuarine habitat took place in the nineteenth and twentieth centuries. Embankments were built along the estuary near the city of London in the nineteenth century. Some of these formed part of the new interceptor sewers commissioned by the Metropolitan Board of Works (Wood, 1982). The new embankments resulted in a loss of both intertidal mudflats and marshland. In addition, some severely polluted and heavily silted freshwater tributaries (e.g. river Fleet) were incorporated into London's sewerage system.

In the 1970s, London's flood defences were upgraded to deal with the threat of tidal surges. The Thames Barrier was constructed at Woolwich and new flood walls were built, or existing walls increased in height, downstream of the barrier. Smaller 'drop-gate' barriers were constructed across the Roding, the Darent and parts of the Canvey Island Creek System, where access by boat was necessary. These barriers were kept open except at times of high flood risk. The mouths of the Ingrebourne, Rom/Beam and Mar Dyke tributaries were closed off by flood defence walls and the freshwater flow from these rivers was ducted to the estuary via a sluice. Figure 2.3 shows the lower reach of the Ingrebourne with the mouth in-filled behind a flood defence wall. The sluice ducting water to the estuary can also be seen.

The earlier conversion of tributaries into part of London's sewerage system and the more recent closing off of tributaries as a part of flood defence improvements represents a significant loss of habitat within the estuary. It is generally accepted that estuaries are important nursery and overwintering ground for many fish species (Claridge *et al.*, 1986; Elliott *et al.*, 1990). In addition, there is evidence to suggest that tributaries can act as a refuge for fish when water quality is poor in the main estuary. Möller and Scholz (1991) showed that during the presence of a low oxygen zone in the Elbe estuary, fish concentrated in a tributary where water quality was much improved. They concluded that the protection of such tributaries was particularly important if major fish mortalities were to be avoided. In the Thames estuary, the Ingrebourne, Rom/Beam and Mar Dyke are in the euryhaline reach of the estuary which has the poorest water quality and currently suffers most from storm discharge events.

2.8 THE CURRENT APPROACH TO ENVIRONMENTAL MANAGEMENT

2.8.1 Water quality management

In 1989, privatization of the water industry took place in England and Wales and the responsibility for water quality monitoring and pollution control in the Thames estuary was transferred to the Thames Region of the newly created National Rivers Authority (NRA). A further change took place in April 1996, when the Environment Agency (EA) was created by combining the NRA, Her Majesty's Inspectorate of Pollution (HMIP) and the numerous Waste Regulation Authorities. Despite these changes the current approach to water quality management of the Thames estuary is in many ways the same as that developed by the Thames Water Authority in the late 1970s and early 1980s. However, the water quality objectives

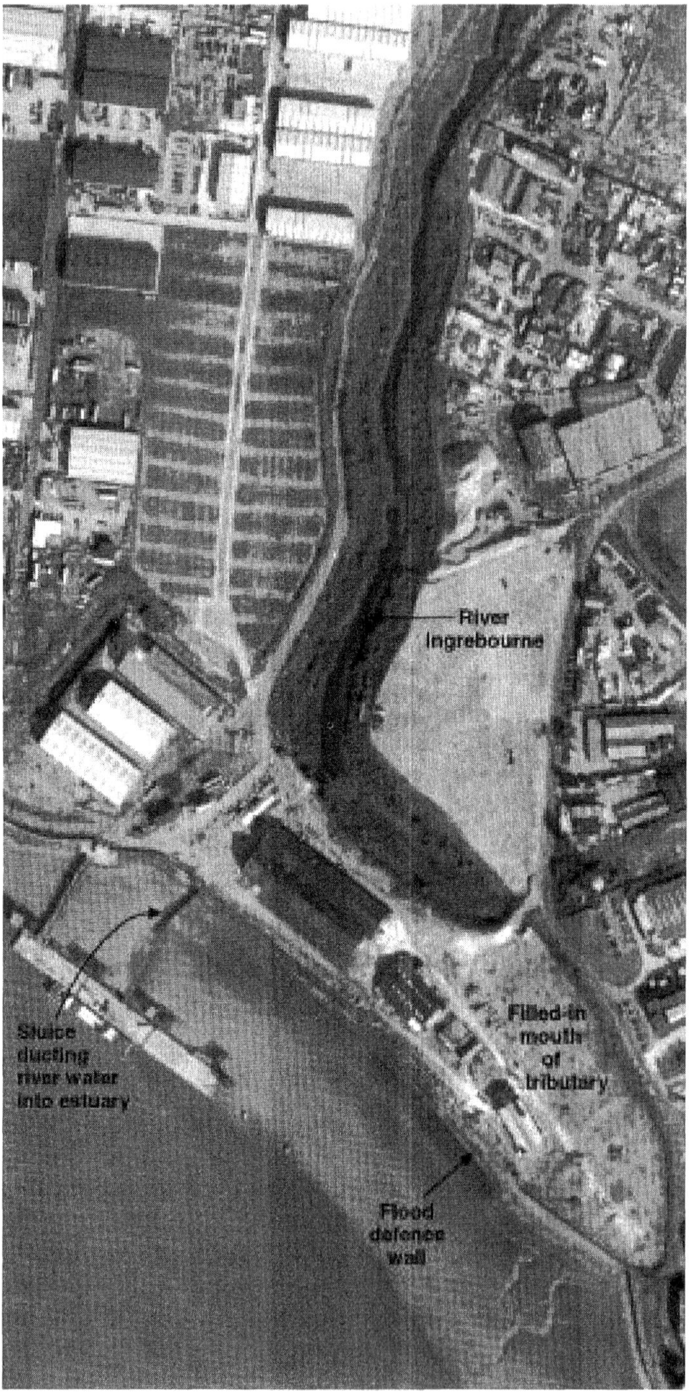

Figure 2.3 The lower reach of the river Ingrebourne.

formulated in the late 1970s have since been revised to represent the present situation more clearly.

The estuary is now divided into three new reaches for the purposes of water quality management. These better reflect the current salinity regime to be found in the different parts of the estuary. A number of use-related objectives and standards have been set for water quality within each reach. The standards reflect both the water quality objectives for that reach and where the quality is better than the use-related objectives demand, the achievable quality at present, so as to guard against any deterioration. A list of these revised objectives and the standards are given in Table 2.3. Compliance with the standards should help to ensure that the use-related water quality objectives are achieved.

An absolute dissolved oxygen standard of 5% saturation has been introduced to provide protection against major fish-kills following storm discharge events. The dissolved oxygen standard of 10% saturation for 95% of the time described in section 2.6.1 has been retained to demonstrate that standards have not been relaxed since the late 1970s. Finally, the dissolved oxygen standards of 40%, 30% and 60% saturation also described in section 2.6.1 for the freshwater, euryhaline and marine reaches, respectively, have been adopted as absolute standards for 80% of the time instead of as 95 percentiles or as mean values. Water quality is exempt from meeting these standards for 20% of the time because experience indicates that estuarine fish populations can tolerate lower oxygen concentrations for short periods. The figure of 20% relates to the current frequency of storm discharges within the estuary.

The temperature standard of 28°C, recommended by the EU for the protection of coarse fisheries, has been adopted to help to protect estuarine fish populations from the lethal effects of increasing water temperatures within the estuary. The freshwater reach of the estuary has the maintenance of such a fishery as one of its use-related water quality objectives. The more stringent temperature standard of 21.5°C, recommended by the EU for the protection of salmonid fisheries, has been used to help protect salmon migration from the effects of proposed warm-water effluents from new power station developments. Monitoring compliance with environmental quality standards for the dangerous substances described in section 2.6.5 has also been included.

Clearly measurable biological standards were included as a guide to water quality. In the freshwater reach, the presence of dace of the year was determined, reproductive success being taken as a sublethal indicator of good water quality. Also in the freshwater reach, the presence of a number of freshwater macroinvertebrates with a known pollution tolerance was used as a measure of water quality. The standard was expressed as a Biological Monitoring Working Party (BMWP) score = 25. In the euryhaline reach, a minimum of nine fish species caught on the cooling water intake screens at West Thurrock Power Station during a four-hour survey

Table 2.3 Environment Agency, Thames Region – estuary water quality objectives and standards

Reach	*Water quality objective*	*Standard*
Teddington to Battersea (freshwater)	• Passable to migratory fish • Maintenance of a coarse fishery • Aesthetically pleasing appearance	• DO > 40% saturation for 80% of the time • DO > 10% saturation for 95% of the time • Minimum DO = 5% saturation • Maximum temperature = 28°C • Compliance with requirements of appropriate EC Directives as laid down in Statutory Instruments • Presence of dace of the year • BMWP > 25
Battersea to Mucking (euryhaline)	• Passable to migratory fish • Maintenance of a euryhaline fish population • Maintenance of a commercial eel fishery • Aesthetically pleasing appearance	• DO > 30% saturation for 80% of the time • DO > 10% saturation for 95% of the time • Minimum DO = 5% saturation • Maximum temperature = 28°C • Compliance with requirements of appropriate EC Directives as laid down in Statutory Instruments • Minimum of 9 fish species caught during West Thurrock Power Station fish surveys
Mucking to Seaward Limit (marine)	• Passable to migratory fish • Maintenance of a marine fishery • EC bathing waters at Southend • Aesthetically pleasing appearance	• DO > 60% saturation for 80% of the time • DO > 10% saturation for 95% of the time • Minimum DO = 5% saturation • Maximum temperature = 28°C • Compliance with requirements of appropriate EC Directives as laid down in Statutory Instruments

DO = dissolved oxygen

was used as a biological standard until 1993, when the power station closed. The location of the power station and details of the survey method are given in Chapter 7. An improved set of biological standards is currently being developed.

2.8.2 Operating agreements between the EA, Thames Region and Thames Water

Operating agreements exists between the Environment Agency and Thames Water covering temporary improvements to the standards of sewage treatment at the main estuary works, the suspension of water abstraction upstream of Teddington Weir and the deployment of the in-river oxygenation unit. Ownership and the cost of maintaining and operating the oxygenation unit has remained with Thames Water after privatization as they are still responsible for the storm discharge problem. However, decisions as to when and where the unit is needed rest with the Agency, as the regulators of water quality within the estuary.

2.8.3 The current approach to habitat management

The most important habitats within the Thames estuary have been recognized and designated for their value. These range from extensive sites of international or national significance to smaller sites of more local importance. The international importance of the estuary for wildlife is recognized by the designation of Benfleet and Southend Marshes as a Special Protection Area (SPA: EC Birds Directive) and RAMSAR site. A number of sites are recognized as being of national importance by designation as Sites of Special Scientific Interest (SSSI) or as National Nature Reserves (NNR). The SSSI is a legal designation made by English Nature under the Wildlife and Countryside Act (1981). The SSSI most widely known for its history of conflict between conservation and development within the greater Thames estuary is that which includes Rainham Marshes. English Nature may own or lease NNR sites or just manage them under a Nature Reserve Agreement (Davidson, 1990). The NNR nearest to London is the one at Leigh, near Southend. Voluntary conservation bodies have also identified areas of local importance within the estuary, but such non-statutory designations offer less protection to estuarine habitats than those made by English Nature.

Opportunities to restore, enhance and create estuarine habitats arise with proposed developments along the Thames estuary. The Environment Agency has a role to play here, particularly in relation to the maintenance and replacement of London's flood defences. In future, consideration could be given to re-opening the mouths of tributaries such as the Ingrebourne, Rom/Beam and Mar Dyke described in section 2.7. Such actions would reinstate creek habitat and as a result increase the value of the estuary as a nursery and overwintering area for fish. As indicated in section 2.7, this type of habitat restoration would also help to protect estuarine fish populations from storm discharge events. Protection from flooding could still be provided by constructing drop-gate barriers across the tributaries.

In addition, consideration could be given to replacing some of the flood defence walls with 'soft engineering' options, particularly where the land behind the wall is no longer being used for agriculture or residential and industrial purposes. Such an approach could reduce the future budgets required for flood defence work and would create valuable estuary habitat (Pearce, 1992).

CHAPTER 3

Physical and chemical characteristics

Jane Kinniburgh

3.1 INTRODUCTION

The ecosystem that the Thames estuary can support is dependent on the physical and chemical characteristics of the environment. This chapter looks at the physical and chemical environment and demonstrates how it varies in time and space. Variations are due to both natural and anthropogenic processes. The variations can be complex and are not understood completely; mathematical models are used to help to gain an understanding of how the system works, and these are described briefly.

3.2 PHYSICAL CHARACTERISTICS

3.2.1 Topography

As stated in the Chapter 2, the Thames estuary is V-shaped with a bell mouth varying in width from 0.1 km at Teddington to 7 km at Southend-on-Sea (widths at average tidal level). With an approximate total length of 110 km, it is one of the longest estuaries in Britain. Figure 3.1 shows the relationship between depth and distance, the data being derived from hydrographical surveys. The data relate to average tidal levels and both widths and depths will be greater at high tide than low tide, and on spring tides. Average depths range from 2 m in the upper estuary to 10 m in the outer estuary. Figure 3.1 also shows cross-sections at London Bridge, Tilbury and Southend. These show that depths vary significantly across the estuary; for example, the average depth at Tilbury is 10 m but the maximum depth is 19 m.

A Rehabilitated Estuarine Ecosystem. Edited by Martin J. Attrill.
Published in 1998 by Kluwer Academic Publishers, London. ISBN 0 412 49680 1.

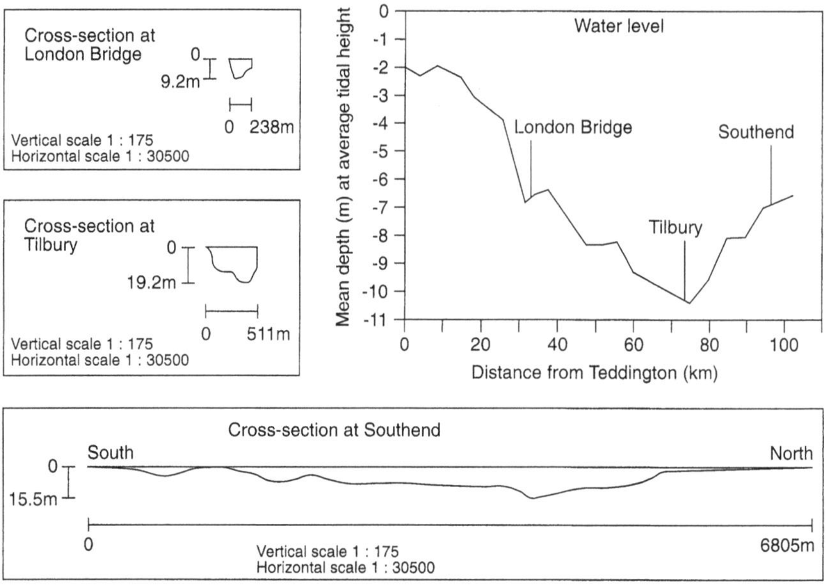

Figure 3.1 Depth profile and cross-sections of Thames estuary.

3.2.2 Water levels

Figure 3.2 shows the typical monthly variation in water level due to tides at Southend (100 km from Teddington) and at Chelsea (22 km from Teddington) for May 1991. The data are derived from a model of the Thames estuary which was originally calibrated against data from tide gauges. The figure shows clearly the two daily tidal cycles, with bi-weekly cycles superimposed. Each day there are two high tides and two low tides, due to the effect of the moon's gravitational field on the earth. The bi-weekly cycles are due to the effects of the sun and the moon which lead to two spring tides (maximum difference) and two neap tides (minimum difference) each month. Spring tides cause daily water levels to vary by 6 m at Southend; neap tides by 3 m. Due to the relative positions of the three celestial bodies, one spring tide each month is greater than the other (in this example, in the middle of the month). There are other regular annual oscillations imposed on this pattern, e.g. the highest spring tides occur at the equinox, and there are other longer period (several years) oscillations. However, these effects are very small relative to the diurnal and monthly events.

In reality, the water levels measured will vary from those shown in the figure because of the effects of freshwater flow and the weather, particularly the wind. These cause variations ranging from a few centimetres up to a metre.

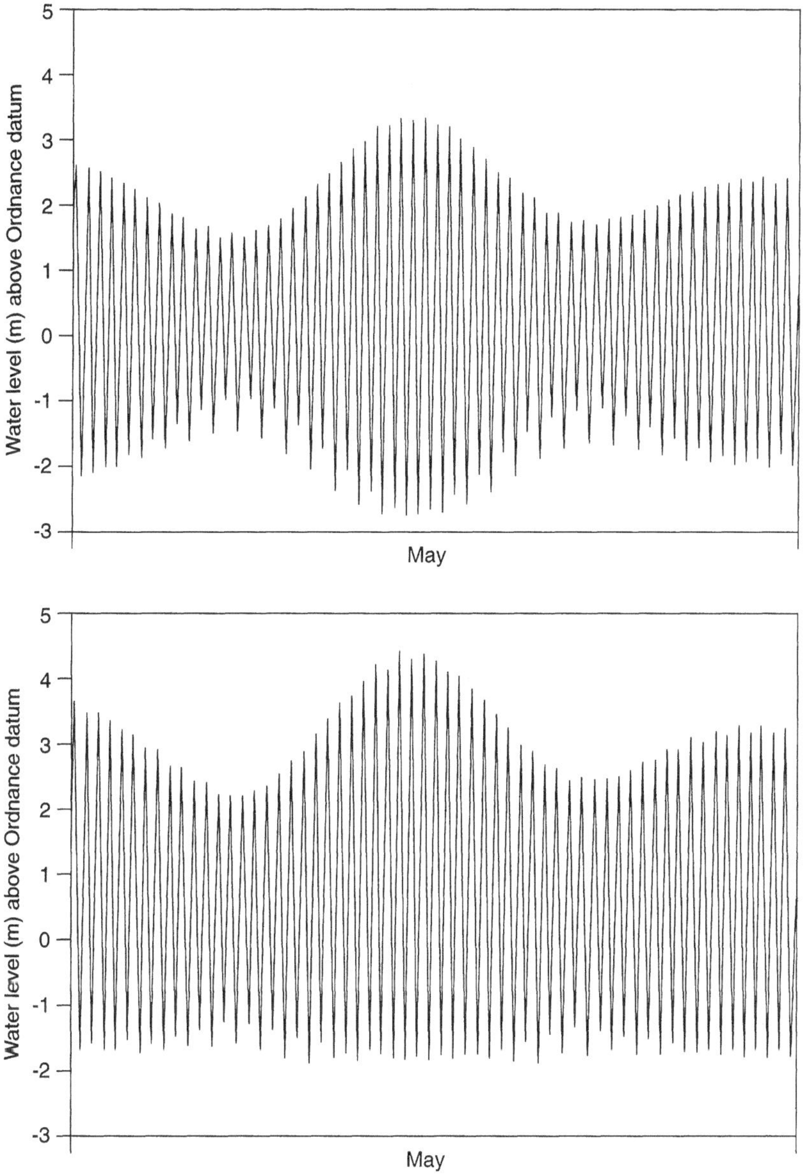

Figure 3.2 Water levels during May 1991 at **(a)** Southend and **(b)** Chelsea.

Figure 3.2 shows that high tides reach about 4 m above Ordnance datum in the upper estuary, and only just over 3 m in the outer estuary (in this example). The difference is due to the effects of the shape of the estuary on water movements.

The figure for Chelsea shows a truncation of levels in the low tides. This is due to the effects of a high-tide lock at Richmond which helps to maintain levels at low tides.

3.2.3 Water movements

As well as affecting the water levels, tides create the largest movement of water in the estuary. The average distance that the water moves (tidal excursion) during a cycle is 12–15 km. The movement creates currents in the estuary, which are faster at spring tide than neap tide, and faster mid-river than at the banks. Velocities are greatest when the water is moving in (flood tide) or running out (ebb tide) with a period of very low velocity (slack water) at the times of high tide and low tide. Velocities vary with position in the estuary, due to frictional effects, topography and shape, but are in the order of 1 m/s during the flood and ebb. During spring tides, the velocity increases to 1.4 m/s but on neap tides only reaches a maximum of 0.6 m/s.

As well as determining currents, the tidal movements are important in determining the quality of the water. For example, any pollutant entering the estuary at a point will be moved upstream and downstream by these movements. This will occur simultaneously with dispersion in the water body. To a certain extent, the water movements help in the mixing of pollutants. However, compared with rivers where flow is in one direction, pollutants will remain in the estuary for a longer time. The retention (or residence) time (i.e. the time that a pollutant remains in the estuary) depends on the amount of freshwater flow entering the estuary causing a net movement of water seawards. At low freshwater flow, water in the upper estuary is displaced seawards at a rate of 3 km per day, but this increases to 11 km per day at average flows. Similar figures for Tilbury are 0.3 and 1 km per day, respectively, slower due to the greater mixing volume there. The average time taken for water to pass from Teddington to the outer estuary is about one month at average freshwater flows, and three months at time of low flow. Tidal currents are also important in that they cause disturbances to the bed of the estuary. This results in resuspension of bed material which is greater at spring tides because of the greater velocities at that time. These aspects are considered in more detail in section 3.2.6.

3.2.4 Freshwater flow

The previous section showed how water movements in the estuary are affected by freshwater inputs; these are now described in more detail. Table 3.1 lists average flows (exceeding 0.1 cumecs) into the estuary from rivers and large sewage treatment works for the period 1978 to 1988. The freshwater Thames is the largest input, contributing 60% of the total,

Table 3.1 Freshwater flows into the Thames estuary (mean daily flows in cumecs)

Rivers	*Average 1978–1988*	*Low flows July to Sept 1991*
River Thames	67.8	11.2
River Crane	0.6	0.4
River Brent	1.1	0.6
Beverley Brook	0.6	0.5
River Wandle	2.1	1.7
Ravensbourne	0.4	0.3
River Quaggy	0.1	0.1
River Lee	6.9	3.4
River Roding	2.2	0.7
River Beam	0.3	0.2
Ingrebourne	0.4	0.2
River Cray	0.6	0.4
River Darent	0.5	0.1
Sewage Works		
Beckton	12.4	10.6
Crossness	7.3	6.6
Riverside	1.4	1.4
Long Reach	2.2	1.6
Kew	0.5	0.5
Mogden	5.4	5.4
Southend	0.5	0.5
Others	1.0	1.0
Sub-total rivers	83.6	19.8
Sub-total sewage works	30.7	27.6
Total	114.3	47.4

followed by Beckton Sewage Treatment works, contributing 11% of the total. The only other river to make any significant contribution is the River Lee (6%). Some of the other sewage works are large contributors, especially Mogden in the upper estuary and Crossness in the middle estuary, but those seawards of Purfleet are smaller and only contribute 1.5% of the total flow. The sewage works in total contribute 27% of the freshwater flow at average flows. However, at low river flows, the flows provided by the sewage works (58%) exceed those of the rivers (42%). Example data are given in Table 3.1 for the period July to September 1991, a time of low flows. During this period river flows were only 24% of their average.

Flow is an important factor controlling water quality in the estuary. At times of low flow, retention times increase and pollutants will remain longer in the estuary, during which time they may decay (broken down by bacteria), a process that uses up oxygen in the water. At times of high flow, materials (including pollutants) are flushed out. Because the flow over

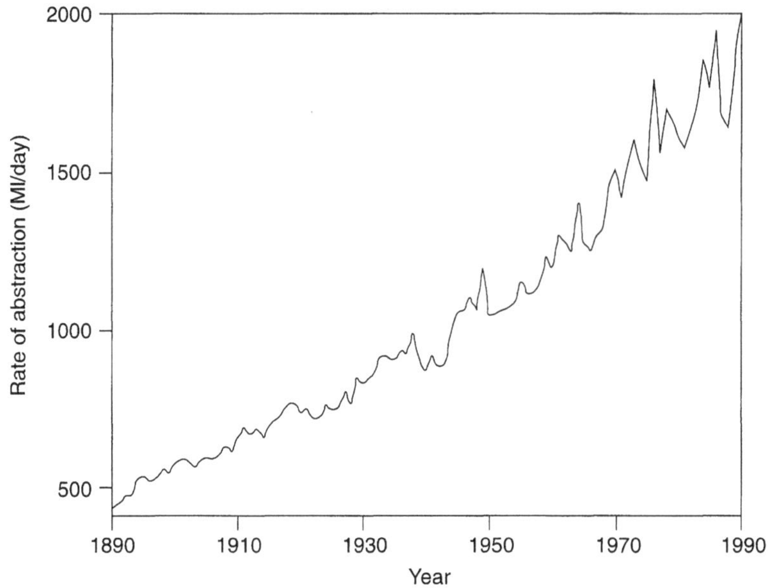

Figure 3.3 Annual abstraction from the lower Thames.

Teddington is such a large proportion of the total flow, it is useful to understand why the freshwater flow in the Thames varies so much.

Between Windsor and Teddington, over a distance of 40 km, there are 11 abstraction points for public water supplies, operated by three water companies. The amount of abstraction is limited by licences issued under the Water Resources Acts. Figure 3.3 shows how the quantity abstracted over the last century has been steadily increasing, from 500 Ml/day in the early twentieth century to 2000 Ml/day currently. This reflects the increased demand for water in the Greater London and Home Counties areas. Constraints are placed on Thames Water Utilities (the largest of the three water companies) on the quantities that can be abstracted at any time during the year in an effort to maintain some flow over Teddington Weir. Normally, the flow over Teddington weir must be 9.3 cumecs (800 Ml/day) but this can be reduced in stages to 2.3 cumecs (200 Ml/day) if their storage reservoirs are lower than threshold levels. Full details are given in Sexton (1988) and Glenny and Kinniburgh (1991).

The impact of these abstractions on river flow is shown in Figure 3.4, which is a plot of Teddington flows for the period 1980–1992. The figure includes a plot of gauged flows, i.e. measured at an ultrasonic gauging station at Kingston, and a plot of natural flows, i.e. gauged flows plus the measured quantities abstracted by the water companies. The figure shows the large variation in flow from year to year and that summer flows are

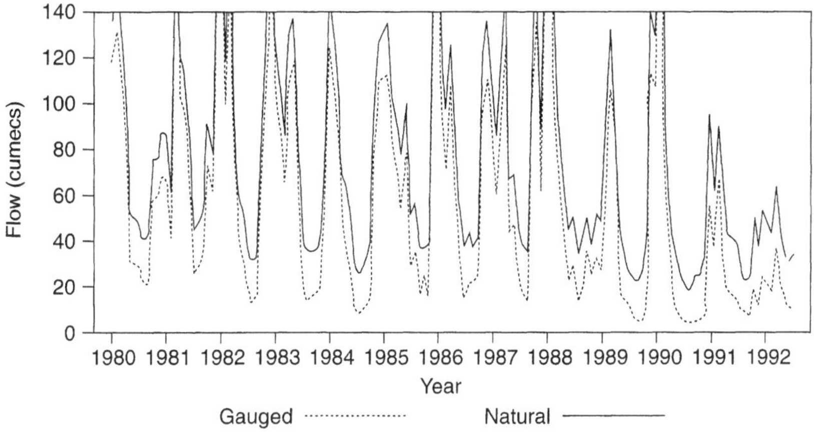

Figure 3.4 Monthly mean flows at Teddington, 1980–1992.

significantly smaller than winter flows. The lowest flows occurred naturally in 1984, 1989, 1990, 1991 and 1992. These are also reflected in the gauged flows. It is interesting to note that the London area's demand for water is approximately 2000 Ml/day (23 cumecs) which exceeds natural river flow in the summer – hence the need for storage reservoirs. During droughts (e.g. 1989 to 1992), the freshwater input from Mogden sewage treatment works, 5 km below Teddington, can be twice as large as the Thames river flow. This has implications for the water quality in the upper estuary since there is less dilution water available for the effluent as well as increased retention times. The effects of flow on water quality are shown in the section on chemical characteristics (section 3.3).

There are a few other freshwater inputs to the estuary from industries. These are mainly paper mills contributing approximately 0.5 cumecs. Other industries (e.g. oil and sugar refineries) do discharge effluents to the estuary but these are predominantly cooling waters which have been initially abstracted from the estuary, and so are not additional freshwater inputs.

3.2.5 Temperature

Water temperatures vary seasonally, typically from 4–5°C in the winter to 21–22°C in the summer. Figure 3.5 shows temperatures in the middle estuary (Beckton) from 1980 to 1992, measured on regular boat runs along the estuary. The data show that the change from winter to summer temperatures (and vice versa) is relatively rapid, generally over a period of a few weeks. Summer temperatures were particularly high in 1983, which is thought to be due to the impact of high output from Littlebrook Power

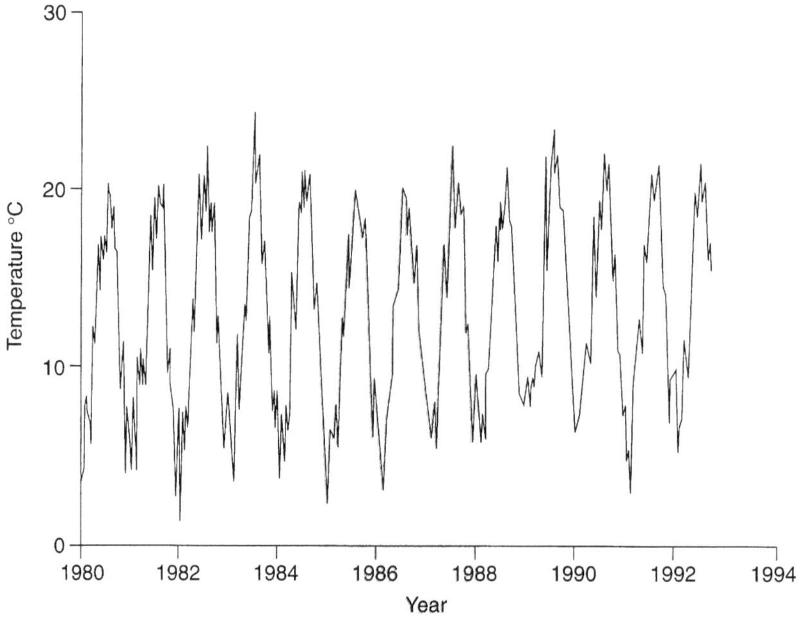

Figure 3.5 Temperatures in the middle estuary (50 km below Teddington), from 1980 to 1992.

Station (60 km below Teddington) at that time. There were also high temperatures in 1989, in both summer and winter (relative to other years), which reflects atmospheric conditions. Reduced freshwater flows are likely to have some impact on temperatures since they will reflect air temperatures more rapidly than the sea will. This could explain the relatively higher winter temperatures in recent years, i.e. the lack of cooler freshwater inputs.

Temperature variations at a point are much greater than over the length of the estuary, and average annual temperatures range from 13.5°C in the upper estuary to 12°C in the outer estuary. Local variations occur because of cooling water discharges associated with power stations, existing ones being at Lots Road (Chelsea), Littlebrook and Tilbury. Others are being planned. Typically, they raise the overall temperatures in the reach of the discharge by about 1°C but locally there are thermal plumes with temperature increases of up to 10°C. These will tend to be more intense in the surface layers due to the buoyancy of warmer water, and their extent will be very much dependent on tidal movements. Furthermore, the impact of a cooling water discharge is less marked as the volume of the estuary increases, which explains why temperatures in the Tilbury area are less affected by a thermal discharge than in the Littlebrook (middle estuary) area.

Temperature has a significant effect on some estuary processes. The lowering of water temperature by 20°C increases the viscosity by a factor of 1.4. This slows down the deposition rate of suspended solids, which means that they remain in the water column longer in the winter than in the summer. Temperature also affects the biochemical processes in the estuary, whereby organic loads are reduced by bacteria. The breakdown is much slower in the winter than in the summer months, reduced effectively to zero at temperatures less than 5°C. Carbonaceous oxidation doubles for a 15°C rise in temperature, and nitrification (oxidation of ammoniacal nitrogen) doubles for a 10°C rise in temperature. This has a significant effect on the oxygen balance, since the bacteria using the oxygen are much more active in the summer than in winter. The solubility of oxygen in water is also affected by temperature (less soluble at high temperatures) so that less oxygen is available to aquatic life in the summer than winter.

3.2.6 Suspended solids

In the previous chapter, turbidity (suspended solids) was stated as an important factor affecting the distribution of estuarine organisms. Suspended solids enter the estuary via rivers, where most of the solids are inorganic or contain partially broken down organic material, and via discharges, which are mainly organic. The suspended solids are moved around by tidal movement but will settle out on the bed when critical velocities (setting velocities) are reached. Because of the varying velocities throughout a tidal cycle, resuspension and settling of solids will take place in a similar cyclical manner. This gives rise to a marked vertical gradient in suspended solids concentrations. Spring tides will resuspend greater quantities of solids than neap tides because of the greater velocities. These processes mean that sediments become mixed, and that organic material deposited on the bed is reactivated when resuspended. Sediments tend to be anoxic, so that resuspension of the sediments can release poor quality water into the column above, creating an oxygen demand. Furthermore, organic material is more easily resuspended (being less dense) than inorganic material and this will add to the oxygen demand.

Trace metals change from being in solution to attaching themselves (partitioning) to particulates. This tends to occur most rapidly in the zone of the highest suspended solids (turbidity maximum). This means that metals with a high affinity for particles (e.g. cadmium) will be depleted from solution in this zone. Once metals are deposited on the bed, then the lack of oxygen may cause them to change from the particulate state to the dissolved state again. If the sediment is resuspended, then the dissolved species will again be released and the metals oxidized.

It is difficult to get an idea of the distribution of suspended solids in the Thames estuary because the processes outlined above mean that there is a great variation in suspended solids throughout the tidal cycle, as well as

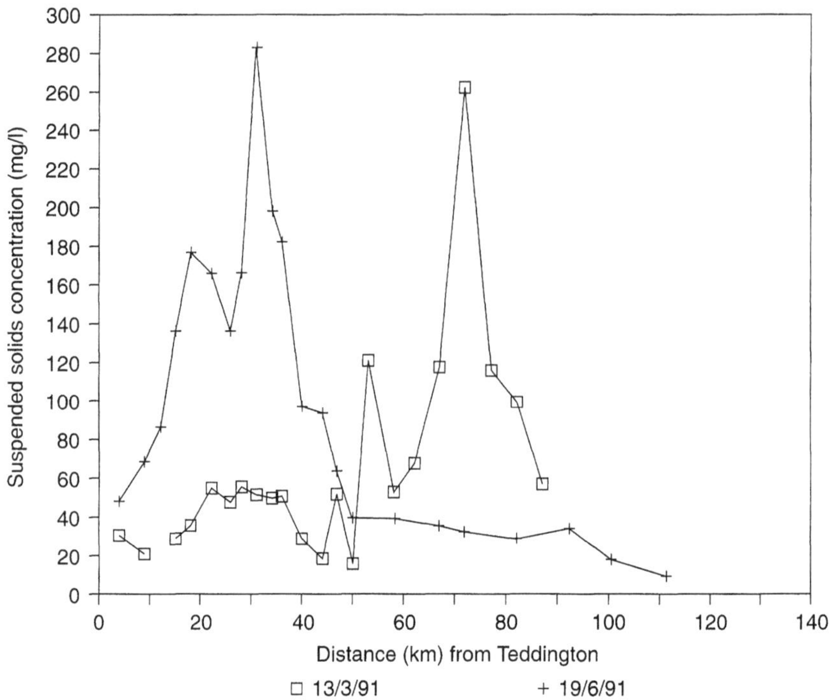

Figure 3.6 Suspended solids concentrations profiles.

there being variation from place to place. Data are collected frequently by the Environment Agency on their boat runs, but the results from place to place will depend on the state of the tide when the sample was taken. Large changes can occur at a point in a few hours. Figure 3.6 gives the results from two boats runs in 1990. The profile for 13 March shows the maximum concentrations in the middle estuary but quite a different profile was obtained in June. These data cannot be used to compare locations because of the impact of tidal state on the data, but they do show the order of magnitude of the suspended solids concentrations (50–300 mg/l) in the surface water layers. The concentrations near to the bed are likely to be 10 times as great. Data given in Technical Paper 11 (HMSO, 1964) record a maximum of 18 928 mg/l at a depth of 14 m on the occasion of a particularly high tide, 50 km below Teddington, but this was exceptional. The paper also gives other examples of suspended solids variations with location and tidal state.

Surveys over a tidal cycle are required to gain a better understanding. Inglis and Allen (1957) showed that particulates accumulate 40–55 km below Teddington, in the reaches commonly known as the 'mud reaches'. Odd (1988) showed that this maximum occurs independent of flow conditions.

Theoretically, one might expect the maximum to occur at the point of zero net flow so this should move upriver with lower flows. However, Odd suggests that this does not necessarily appear to be the case in the Thames estuary. He considers that this is related to rapidly rising bed levels in the upper estuary which cause a significant increase in the slope of the mean tide level, and hence affects processes. Odd also showed that fine bed sediment accumulates in the same area (the mud reaches) though Inglis and Allen (1957) suggested an additional region 70–80 km from Teddington, near Gravesend.

3.3 CHEMICAL CHARACTERISTICS

3.3.1 Salinity

The estuarine environment is significantly different from both freshwater and marine environments because of the variable salt concentrations. Salinity varies from being similar to freshwater in the upper estuary to being similar to sea water in the outer estuary. At any location, salinity varies daily with the state of the tide and seasonally with the amount of freshwater flow. Salinity is not measured directly by the Environment Agency, but chloride is determined on samples taken regularly from the estuary. Salinity, expressed in parts per thousand (i.e. PSU, g/l or ‰), can be calculated from chloride concentrations (mg/l) by multiplying by 0.0018065 (Wooster *et al.*, 1969). Figure 3.7 shows chloride concentrations in the Thames estuary for two sampling trips done from a boat in 1990, one in January and one in August. These data have been half-tide corrected, i.e. measurements made are plotted at the distance where the water would be at half-tide (volume of water to landward is equal to its average value over an average tidal cycle). This is important for comparisons to be valid and to remove the effects of tidal movement.

Figure 3.7 shows how the profile varies with the time of year due to the impact of flow on salinity: high chloride concentrations are further up the estuary during periods of reduced flow (August). The variation at the seaward end is less marked. The figure also shows how the salinity of the upper 30 km is always fairly similar to freshwater (a chloride concentration of 5000 mg/l is equivalent to a salinity of 9 PSU) and the lower 30 km is similar to saltwater (a chloride concentration of 19000 mg/l is equivalent to a salinity of 34 PSU) with an euryhaline zone in the middle. As well as varying with distance, chloride concentration at a location varies with tidal state – as shown in Figure 3.8, which shows chloride concentrations calculated from conductivity, measured at an automatic monitoring station at Greenwich for two months in 1991. It shows that the daily chloride concentration variation due to the tides at this point is about 2500 mg/l. The impact of the spring/neap tide cycle is not that apparent but the effect

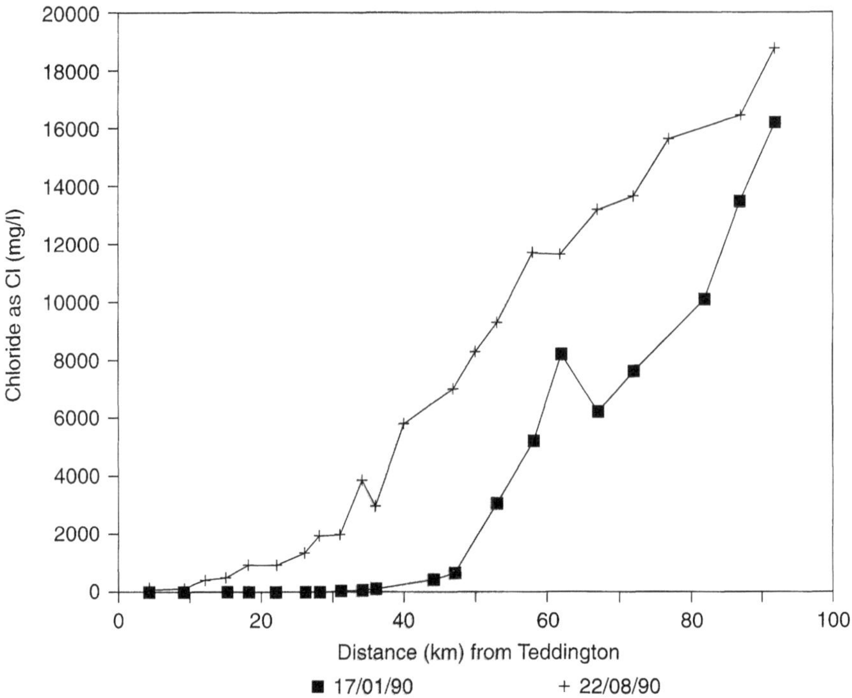

Figure 3.7 Profiles of chloride concentrations, summer and winter in the Thames estuary.

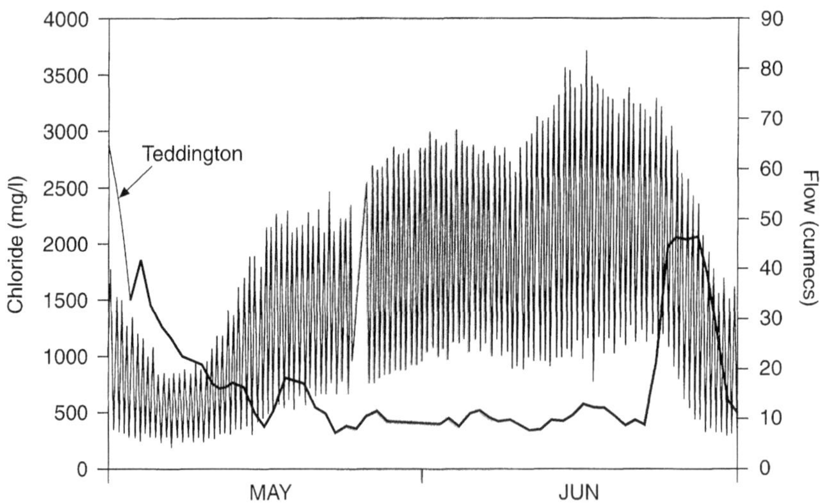

Figure 3.8 Chloride variations over two months in 1991 as measured at the automatic monitoring station at Greenwich. (Flows at Teddington are also shown.)

of flow is more significant. Tidal variation in chloride concentrations is smaller when flows are higher (early May on the figure) and chloride concentrations are greater with lower flows at Teddington.

Section 3.2.4 demonstrated the variation in freshwater flow over the period 1980–1992. This is reflected in chloride concentrations over the same time period, as shown in Figure 3.9 for London Bridge. The higher concentrations each summer are clear, as are the increased concentrations in years of low flow (1983, 1984, 1989, 1990 and 1991). Figure 3.10 shows the effect of low flows in the whole of the upper estuary. This is a three-dimensional plot of chloride concentrations from Teddington to London Bridge for the time period 1980 to date, plotting annual averages so that seasonal variability is excluded. The low flow years cause elevated chloride concentrations. Technical Paper 11 (HMSO, 1964) has shown that the Thames estuary is well mixed vertically, but the paper also shows that, during the run of the tides, there are measurable differences due to the most rapid currents being near the mid-stream. At slack water, though, there is little difference between salinity at the surface and near the bed, or between salinity in mid-stream and near the banks.

3.3.2 Dissolved oxygen

Adequate dissolved oxygen is one of the most important factors in ensuring a thriving estuarine environment. Dissolved oxygen can vary significantly during the day, so in order to monitor concentrations there are eight automatic monitoring stations in the Thames estuary, between Kew and Purfleet, operated by the Environment Agency. They are located so as to cover the whole of the tidal cycle in the upper and middle estuary. Figure 3.11 shows typical dissolved oxygen profiles as measured by the automatic monitoring stations (AQMS) for winter and summer, respectively. They have been simplified since there are four traces per day, corresponding with the number of tidal movements. There is a characteristic 'sag' in dissolved oxygen about 50–60 km below Teddington. This is due to the impact of the discharges from Beckton and Crossness sewage works. Even though the effluent is treated to a high standard, because the flows are large the polluting load causes a drop in dissolved oxygen as bacteria use this up in the breakdown of the polluting load. In the summer an upriver sag may also occur. This is due to the load from Mogden sewage treatment works and sporadic influence of discharges from storm drains. The sag does not occur in the winter when flows are high, the retention time is short, and the breakdown of the polluting load is slowed down by the lower temperatures. However, with low summer flows this load remains in the vicinity longer, exerting an oxygen demand. The figure also shows that, overall, dissolved oxygen concentrations tend to be higher in the winter than in the summer. This is due to the effect of flows and temperature. As stated in a previous section, oxygen is less soluble at

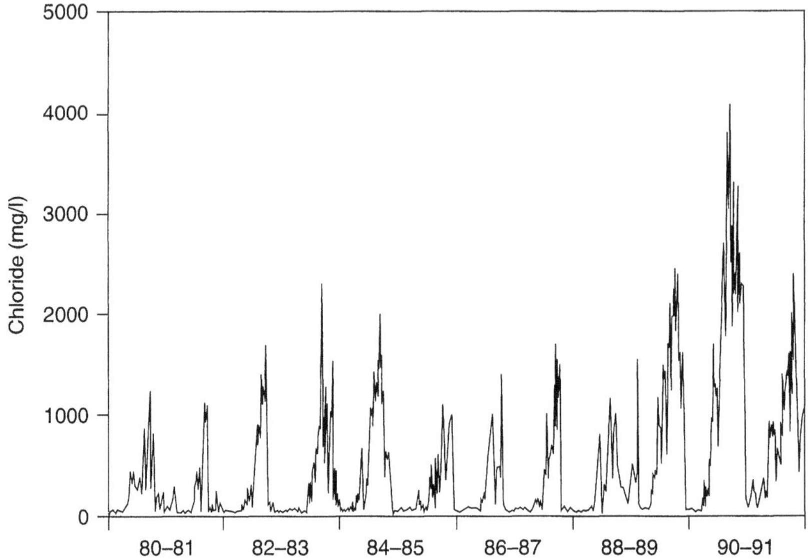

Figure 3.9 Chloride concentrations at London Bridge, 1980–1991.

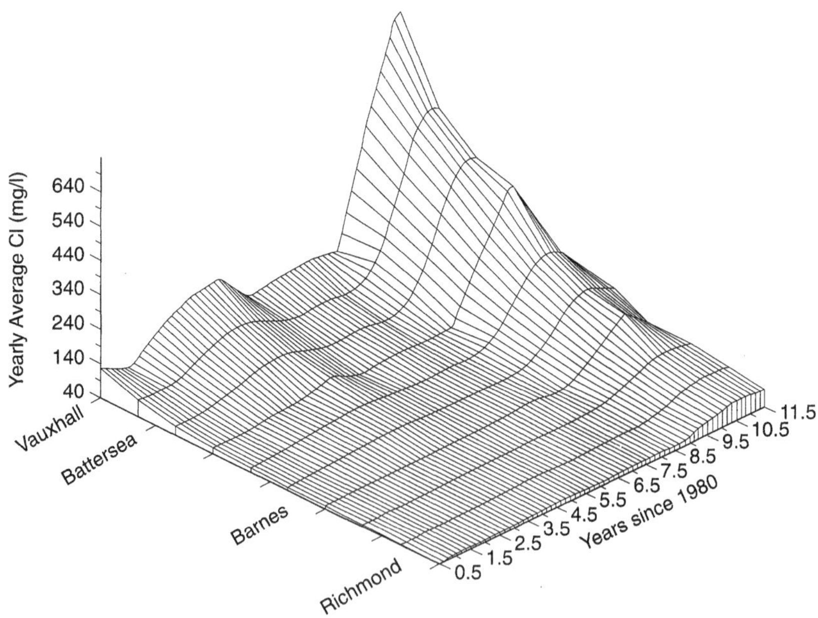

Figure 3.10 Average annual chloride concentrations in the upper estuary, 1980–1991.

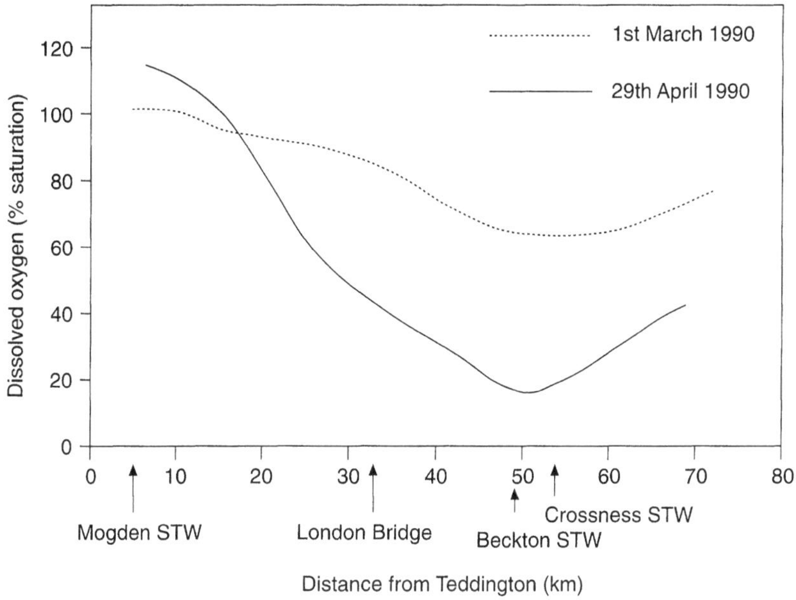

Figure 3.11 Dissolved oxygen profiles in summer and winter.

higher temperature and bacterial activity in the breakdown of polluting material is more rapid at higher temperatures, hence using up more of the available oxygen.

In order to explain the variation in dissolved oxygen with time, the quality of the effluents discharged needs to be considered, since this can vary daily and from discharge to discharge. The quality is usually determined in terms of biochemical oxygen demand (BOD) and ammoniacal nitrogen. Account must also be taken of the quantity of flow to determine the effective oxygen load (EOL) on the estuary. This has often been assumed, for comparative purposes, to be:

$$EOL = flow \times [(1.5 \times BOD) + (4.5 \times N)]/1000$$

where EOL is the effective oxygen load in tonnes per day, BOD (mg/l) is measured over five days in the presence of allyl thiourea to suppress nitrification (i.e. BOD_5), N is organic nitrogen plus ammoniacal nitrogen (mg/l), and flow is in Ml/day.

Table 3.2 gives example EOLs for discharges to the estuary. The table shows that the organic load from the sewage works is far greater than from any other source.

Dissolved oxygen variations with time are shown in Figure 3.12 for three locations in the Thames estuary: Kew (upper estuary), Beckton (middle estuary) and Coryton (outer estuary). The seasonal cycle is

Table 3.2 Relative pollution loads of discharges to the Thames estuary

Type of discharge	*Effective oxygen load (tonnes/day)*
Major sewage treatment works	120
Small sewage treatment works	45
Sugar refineries	1
Paper mills	10
Oil refineries	8

pronounced at Kew and Beckton but not apparent at Coryton in the outer estuary. At Kew, highest dissolved oxygen concentrations occur in the spring, with a sharp decline in early summer. Phytoplankton living in the water add oxygen during the spring (daytime photosynthesis), causing elevated concentrations of dissolved oxygen with supersaturation (up to 140%) in some years. Supersaturation does not occur further down the estuary because large numbers of phytoplankton cannot exist in the more saline turbid waters of the middle and outer estuary. The plot for Kew shows lower summer minima in recent years. This is probably due to lower freshwater flows.

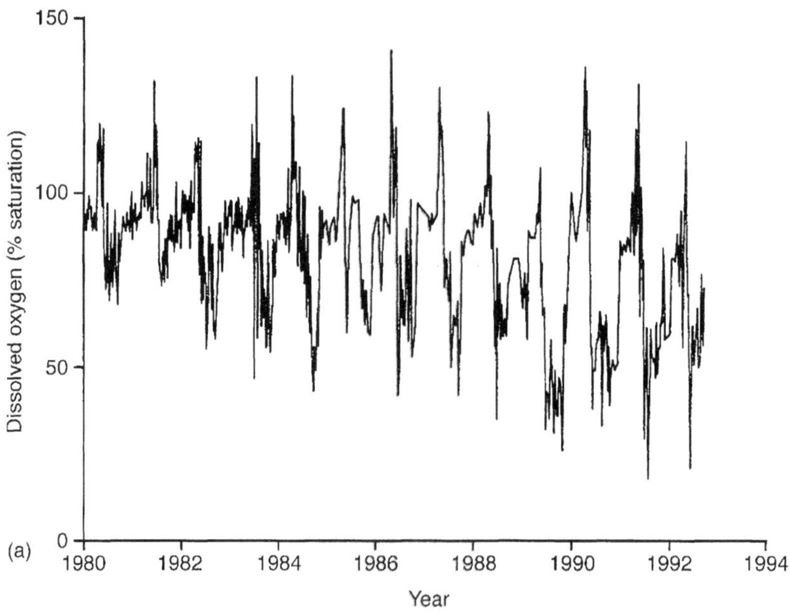

Figure 3.12 Dissolved oxygen concentrations from 1980 to 1992 at (a) Kew, (b) Beckton and (c) Coryton.

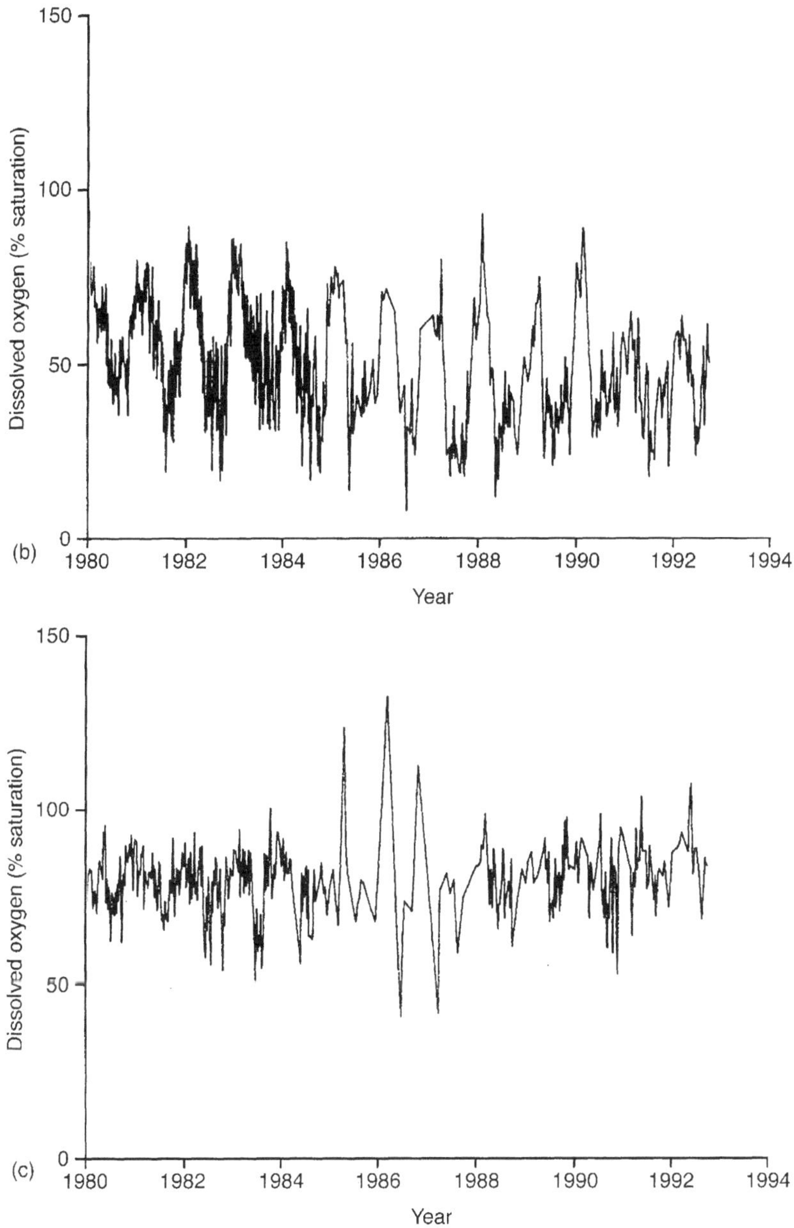

Figure 3.12 (*continued*) Dissolved oxygen concentrations from 1980 to 1992 at (a) Kew, (b) Beckton and (c) Coryton.

Figure 3.13 shows the output from a water quality model which estimates the impact of reduced flows on dissolved oxygen in the estuary. Model runs were done with flows over Teddington of 1600 and

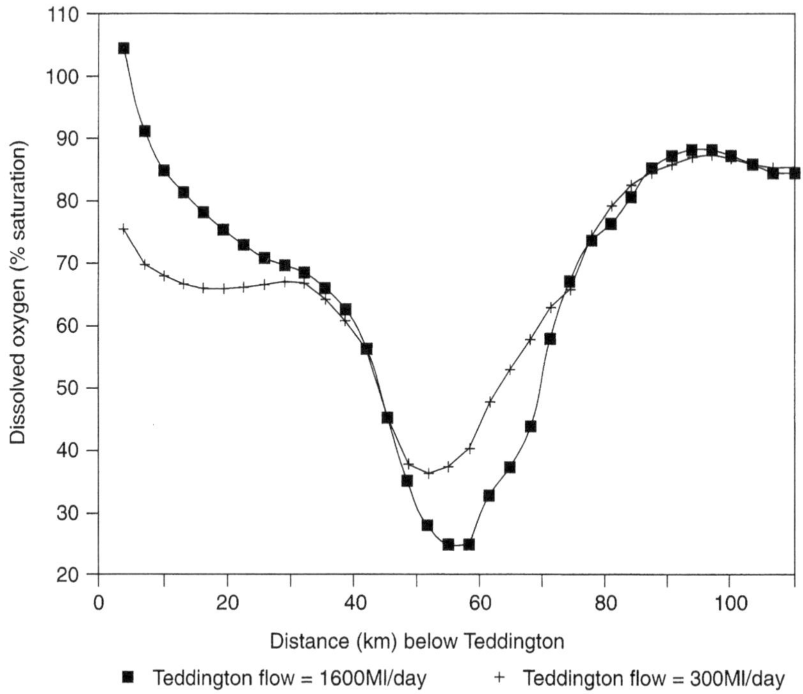

Figure 3.13 Effect of freshwater flow on dissolved oxygen concentrations (model output).

300 Ml/day. All other inputs were left unchanged so that the effect of flow could be shown. The figure shows that reducing the flow by 1300 Ml/day (14.5 cumecs) causes the minimum dissolved oxygen in the upper estuary to be reduced by 10% saturation and the position of the sag minimum to move upstream. This is because of increased retention times of the Mogden effluent. In the middle estuary, a reduction in freshwater flow improves the sag minimum by 10% saturation (less impact of Mogden effluent here). In the outer estuary, flows have no significant effect on dissolved oxygen concentrations. Figure 3.12 shows that minimum dissolved oxygen concentrations at Beckton have improved in recent years compared with the mid-1980s. This agrees with the predicted model output for times of low flow. However, it also reflects recent improvements in the effluent quality from Beckton and Crossness sewage treatment works. In the outer estuary at Coryton, there is less variation in dissolved oxygen. The factors responsible for variations higher up the estuary have less impact in the outer estuary because the discharges have lower loads and the volume of receiving water is greater.

Summer storms can have acute effects on dissolved oxygen concentrations. Figure 3.14 shows the AQMS readings made on two days in 1990,

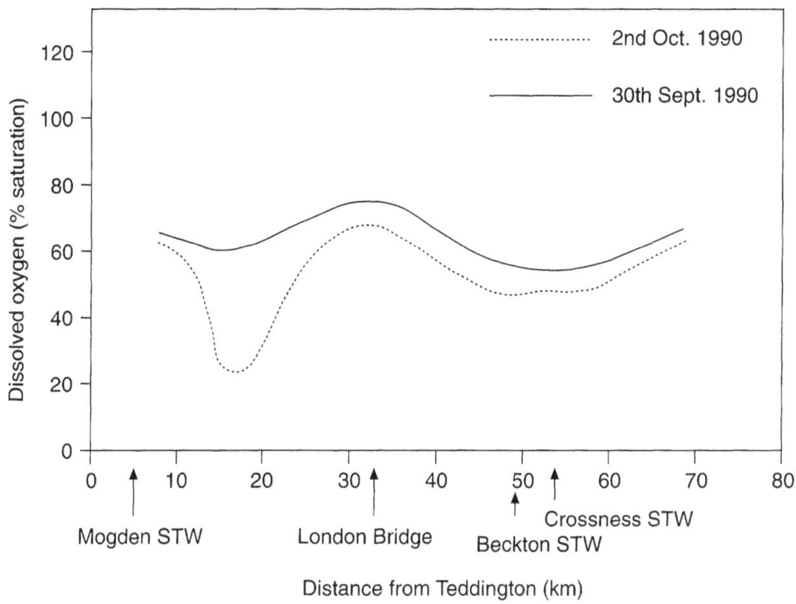

Figure 3.14 Dissolved oxygen profiles during storms.

one prior to a storm on 30 September and one two days later. At times like these, the sewerage system is unable to cope with the volume of flow and direct discharges are made through storm overflows to the estuary. There are over 40 of these between Chelsea and London Bridge, and the quantities discharged can be up to 4000 Ml/day. The EOL for storm discharges is approximately 700 tonnes/day but these loads will only be exerted for a few hours (30 tonnes/hour). The polluting load discharged into the river can cause dissolved oxygen concentrations to fall below 10% saturation, which has a significant impact on aquatic life: fish-kills were recorded in 1986. As mentioned in the previous chapter the Thames Bubbler, which injects 30 tonnes of oxygen per day directly into the affected areas, protects against the consequences of such incidents (Griffiths and Lloyd, 1985).

3.3.3 Nutrients

The availability of nutrients, particularly nitrate and phosphate, is an important factor in determining aquatic life. At Kew, total oxidized nitrogen varies from 5 mg/l in the summer to 14 mg/l in the winter, averaging approximately 10 mg/l. Nitrate concentrations decline seaward, averaging about 2 mg/l at Coryton and 0.5 mg/l at No. 2 Sea Reach (open North Sea). This is due to the fact that the main source of the nitrate is the load from the freshwater Thames catchment, although the sewage treatment works add some nitrate too. Concentrations are highest in the winter because of inputs from surface runoff from agricultural land, the primary source of the nitrate. As nitrate moves down the estuary, it is diluted by increasing volumes of water.

Phosphate, measured as orthophosphate (P), varies between 0.5 mg/l and 5 mg/l in the upper estuary, averaging 2.5 mg/l. The highest concentrations are recorded in the summer months. The main source of the phosphate is sewage discharge in the freshwater Thames catchment. Hence, concentrations in the river are higher in the summer, when the discharges are diluted less than in the winter. Orthophosphate declines seaward to less than 1 mg/l at Coryton and 0.2 mg/l at No. 2 Sea Reach. The concentrations of nitrate and phosphate in the upper estuary are high and are sufficient to support the growth of phytoplankton, provided other conditions are suitable.

3.3.4 Dangerous substances

The presence or absence of toxic substances is an important consideration for the estuary ecosystem. The Environment Agency takes monthly samples at Barnes, London Bridge, Erith, Mucking, Chapman Buoy, and No. 2 Sea Reach for substances listed in the European Dangerous Substances Directives (76/464/EEC) and in the UK Red List. Standards for some of these substances have been set by the Department of the Environment. Table 3.3 gives results for selected substances measured in 1991, together with any Environmental Quality standards that exist, with

Table 3.3 Dangerous substances at London Bridge (annual average)

Substance	*Standard (μg/l)*	*Average 1991 (μg/l)*	*Number samples*
Copper (dissolved)	5.0	9.97	9
Zinc (dissolved)	40.0	44.43	7
Mercury (total)	0.3	0.042	11
HCH (gamma)	0.02	0.018	9
Endrin	0.005	< 0.025	9
Aldrin	0.01	< 0.025	9
Dieldrin	0.01	< 0.025	9
Pentachlorophenol (PCP)	2.0	< 0.4	12
Drins (total)	0.03	< 0.05	9
Cadmium (dissolved)	2.5	0.185	7
Chromium (dissolved)	15.0	3.107	7
Lead (dissolved)	25.0	7.499	7
Carbon tetrachloride	12.0	0.005	11
DDT	0.025	< 0.025	9
Hexachlorobenzene	0.03	< 0.025	8
Hexachlorobutadiene	0.1	0.001	10
Chloroform	12.0	0.4	11

concentrations of metals in sediments being reported by Attrill and Thomas (1995). Many substances are below the detection limit, e.g. dieldrin and pentachlorophenol. Others are well below the Environmental Quality standards. Some substances do exceed the standards, specifically hexachlorocyclohexane (lindane), copper and zinc. Many of these failures are throughout the estuary and not only in the mud reaches, where higher concentrations are expected because of the affinity of some substances for the particles. It is thought that the origins of these substances are in diffuse sources to the sewage treatment works, many of the substances having common domestic uses.

3.4 MODELLING

The previous account shows that the chemical characteristics are often interrelated and depend on the physical characteristics of the estuary. It is often difficult to separate out the effects of one factor because of these relationships. However, mathematical models can be used to do this. Output from such models has been used in the previous sections to give examples of the effect of one factor on another. Models need to take account of hydrodynamics and water chemistry if they are going to simulate estuary conditions effectively. Figure 3.15 shows the chemical cycles that must be included in estuary models and their interrelationships. Models are only as good as the data on which they are calibrated: the previous sections

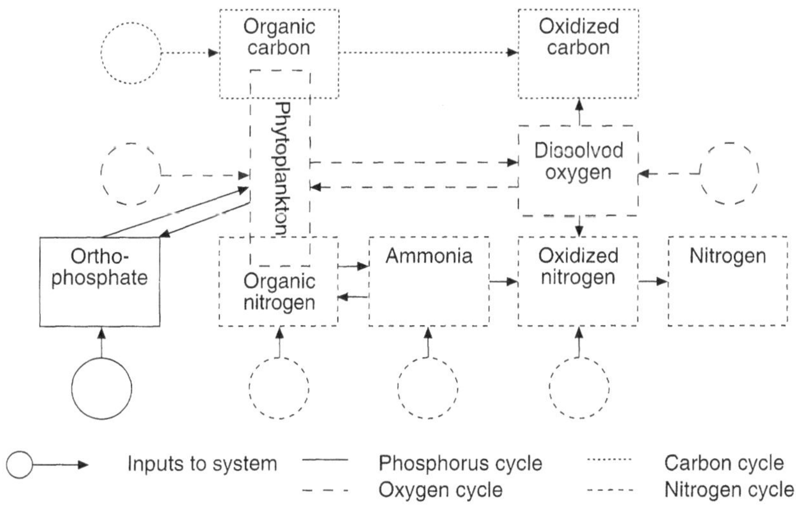

Figure 3.15 Interrelationships in chemical cycles incorporated in Thames estuary models.

show that data are lacking in many areas and that in others the interrelationships are incompletely understood. Much work remains to be done.

3.5 SUMMARY

The physical and chemical characteristics of the Thames estuary have been described and variations in time and space shown. Interrelationships are complex. Some of the most important factors affecting estuary quality are freshwater flow, tidal motion, and the location and size of discharges. Quality variation through time is determined by variation in discharge quality and flow (related to rates of abstraction). The data show that merely considering 'average' conditions may not be that useful ecologically; extremes occur which may have a significant impact on the biota. The drought years of 1989 to 1992 have affected freshwater flows to the estuary and as a result the salinity profiles in the estuary changed temporarily. The situation is complex but the data presented here give an indication of the physical and chemical setting for the Thames estuary ecosystem.

CHAPTER 4

The algae of the Thames estuary: a reappraisal

Ian Tittley and David John

4.1 INTRODUCTION

4.1.1 Background

The benthic marine algal (seaweed) flora of the tidal Thames was first described in detail by Tittley and Price (1977a), prior to which there was only scant reference made to it in the literature. That study was undertaken in connection with a wider appraisal of the seaweed flora of Kent (including the Thames estuary), published as an atlas of species distributions (Tittley and Price, 1977b; Tittley *et al.*, 1985). Since the original study, several other works have added to our knowledge of the algal flora. Zonal and seasonal changes were investigated at Woolwich in the middle reaches of the tidal river (Tittley, 1985); this was the first account of seasonal changes in macroalgal populations for any British estuary. Additional observations on the occurrence and distribution of benthic macroalgae have been made by Price (1982, 1983, unpublished data). The floristic composition, seasonality and distribution of algae in the freshwater reaches of the tidal river were investigated by John *et al.* (1990) as part of a suite of studies on the phytobenthos of the freshwater Thames (John and Moore, 1985a,b; John *et al.*, 1989a,b). The algae and macrophytes of the River Wandle, a south bank tributary that enters the tidal river near Wandsworth, were described by Price and Price (1983).

Studies on the diatom flora of the Thames, in contrast to those on other algal groups, have been much more exhaustive and date back to the mid-nineteenth century. More recently a detailed analysis of the river's diatom communities was undertaken by Juggins (1992). The focus of the investigation was concerned with sediment-inhabiting diatoms and the use of

A Rehabilitated Estuarine Ecosystem. Edited by Martin J. Attrill.
Published in 1998 by Kluwer Academic Publishers, London. ISBN 0 412 49680 1.

such subfossil diatom assemblages to arrive at an estimate of salinity levels at riverside archaeological sites. In that study, little or no attempt was made to distinguish between the epipelic and epipsammic diatoms. Other studies in the Thames estuary also focus on diatom assemblages (e.g. Roper, 1954; Carter, 1933a,b for a saltmarsh on Canvey Island). Greenish patches on mud at mid-tide level in the freshwater tidal Thames reported by John *et al.* (1990) are due to the presence of high surface concentrations of *Euglena*. Information for the estuary on euglenoids and algal groups associated with sediments (other than diatoms) is almost entirely lacking.

Little modern and detailed information exists on the composition, abundance and seasonality of the phytoplankton in the tideway. The most comprehensive studies were those carried out by Rice (1938) in the freshwater tidal estuary over the period 1928–1932, and by Wells (1938) and El-Maghraby (1956) in the outer estuary. A larger survey was conducted for the Thames Water Authority during the late 1970s but no published list is available (Juggins, 1992, pp. 16, 17).

4.1.2 Benthic algal habitats in the tidal estuary

The tidal Thames has undergone many changes since the last glaciation; the more recent of these are as a consequence of human settlement and the river's increasing use as a navigable waterway (Chapter 2). The river is today a much narrower canalized waterway in contrast to its former state, where there existed a wide estuary with extensive marshy embankments. This is confirmed by data associated with early eighteenth century algal records. Dillenius and Ray (1724) and Dillenius (1742) record *Enteromorpha intestinalis* and *Vaucheria dichotoma* occurring in 'marsh ditches' near Charlton, Greenwich and Rotherhithe. River walls at that time also provided habitat: Dillenius and Ray (1724) recorded *Enteromorpha compressa* on river walls at Woolwich.

Since the previous macroalgal survey in the 1970s, a tidal barrage has been built at Silvertown near Woolwich. It is now possible to create temporarily a non-tidal environment between Silvertown and the tidal limit at Teddington; in practice, the barrage is only raised to contain the occasional storm surges or extreme high tide. Associated with the barrage has been the raising or reconstruction of sea-walls along the entire length of the river, and the construction of barrages across the mouth of tributaries such as the Roding at Barking Creek, and the Darent at Crayford (Chapter 2).

The importance of the Thames as a commercial waterway lessened with the removal of port activities from the inner to the outer estuary. This change has not been without its consequences, as new piers and jetties for shipping have provided additional substrata for the attachment and growth of benthic organisms, including algae. Most surfaces available for algal colonization are uncovered during low water and are artificial, generally above mid-tide level and commonly vertical or sloping.

At many sites the extensive foreshore consists of walls and embankments below which are areas of mud, silt and, less frequently, shingle and sand. Recent reconstruction of sea-walls has often resulted in the replacement of the old porous brickwork surfaces by harder impervious cement, with consequences for local algal ecology; other materials commonly present include wood, metal, granite and limestone. Hard surfaces at low levels in the inner estuary are often unsuitable for benthic algae, due to the presence of a layer of diatom-bound glutinous mud; other similarly positioned surfaces descend to low water but remain free of any algae, probably due to very low light penetration during high water caused by extremely high turbidity (cf. Dring, 1984, Chapter 3).

Thameside remains an industrial environment with many factories, power stations, sewage works and other concerns using the river for transportation or as a repository for waste material. The growth of the urban sprawl of London and other towns along the estuary has been, and continues to be, accompanied by the loss of marshland, which is a cause of concern; this also has implications for the associated algal communities. Tracts of marshland still remain in the outer estuary especially saltmarshes dominated by stands of *Puccinellia*, *Spartina* and *Halimione* (these are described more fully in Chapter 9).

Towards the seaward limit of the estuary, extensive foreshore mussel beds and shell or shingle banks provide habitat for algal growth.

4.2 THE PRESENT REAPPRAISAL

This reappraisal will, as did Tittley and Price (1977a), focus on benthic algae other than diatoms, principally the Chlorophyta (green algae), Cyanophyta (blue-green algae), Phaeophyta (brown algae), Rhodophyta (red algae) and Xanthophyta (yellow-green algae), but will draw on diatom information where appropriate.

Its principal aims are to:

- collate all information on the benthic algae associated with hard, stable surfaces in the tidal estuary;
- identify shorter and longer term changes in algal occurrence and distribution (where possible);
- reassess the four floristically defined sections of the river recognised by Tittley and Price (1977a).

4.2.1 Method

Many of the original 34 study sites between Teddington and Shoeburyness (Figure 4.1) were revisited and the macroscopic growths of algae recorded by direct observation. An additional site was investigated at the Thames

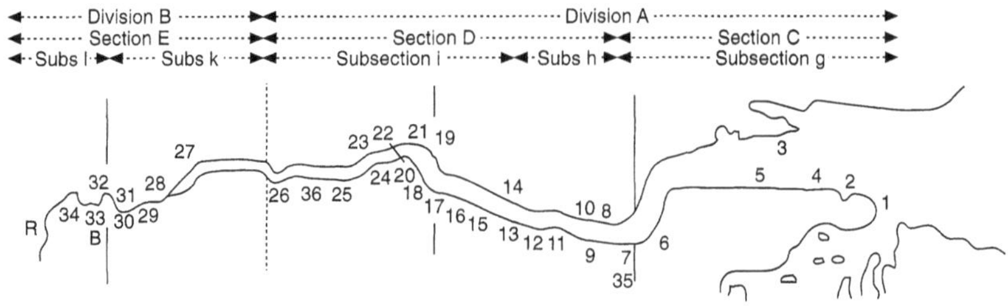

Figure 4.1 The Thames estuary showing sampling sites. Floristic division, sections/subsections from TWINSPAN analysis indicated (Greenwich boundary a hatched line). **1**. Grain; **2**. London Stone; **3**. Canvey Island; **4**. Allhallows; **5**. Egypt Bay; **6**. Shornmead; **7**. Gravesend; **8**. Tilbury; **9**. Northfleet; **10**. Tilbury Dock; **11**. Swanscombe; **12**. Greenhithe; **13**. Littlebrook (Dartford); **14**. Purfleet; **15**. Long Reach; **16**. Crayford Ness; **17**. Erith Marshes; **18**. Erith; **19**. Rainham; **20**. Belvedere; **21**. Dagenham Dock; **22**. Crossness; **23**. Barking; **24**. Tripcock Ness; **25**. Woolwich; **26**. Greenwich; **27**. Central London (Cleopatra's Needle); **28**. Chelsea Bridge; **29**. Battersea Bridge; **30**. Putney Bridge; **31**. Nr Barn Elms; **32**. Hammersmith Bridge; **33**. Barnes; **34**. Kew; **35**. Gravesend Pier; **36**. Thames Barrier. **B** = Barnes (freshwater study); **R** = Richmond (freshwater study).

Barrier, Silvertown. Most green and other small algae were brought back alive to the laboratory for identification.

In the freshwater tidal estuary where the diversity of benthic macroalgae is low, benthic microalgae are the more important component of the flora. Since the benthic microflora is difficult to collect and is usually overlooked in surveys of rivers and their estuaries, these algae were sampled by means of artificial settlement surfaces placed into the field. Polystyrene Petri dishes and polyethylene bags were fixed to suitable wooden structures, at 0.5–4 m above Ordnance datum, and left for about 28 days. Three sites (two at Richmond, one at Barnes) were sampled in this way. The surfaces were subsequently brought back to the laboratory for direct microscopic examination and identification of the colonizing algae.

A database of the distribution of marine macroalgae in the estuary was prepared and, unlike in the first study, was subjected to numerical analysis to test the null hypothesis that there are no floristically distinct regions in the estuary. Three methods were used:

1. A gradient analysis using ordination (DECORANA, detrended correspondence analysis, Vespan II package, Malloch, 1988).
2. A classificatory approach using ordination analysis (TWINSPAN, two-way indicator species method, Vespan II package, Malloch, 1988).
3. An ecocladistic method (PAUP version 2.4, Swafford, 1985; see also Lambshead and Paterson, 1986).

The outputs from each of these three methods are:

- for (1), an ordination plot or scattergram of points representing sites;
- for (2), a dendrogram, the branches of which usually represent sites as either individuals or groups;
- for (3), an ecocladogram, the branches of which represent individual sites.

Ecocladograms are presented in two forms: a geometrically symmetrical diagram; or an asymmetrical diagram in which branch lengths are scaled according to the number of attributes (species) separating the nodes or characterizing the terminal branches.

4.3 THE ALGAL FLORA BY HABITAT

4.3.1 Marine to brackish-water reaches

The algal vegetation colonizing river-walls, mussel shingle and shell banks, saltmarshes and floating structures remains essentially similar to that described by Tittley and Price (1977a). Recent reappraisal at Gravesend, for example, revealed a band of algal vegetation at around high tide level dominated by extensive growths of the small green alga *Blidingia minima;* this algal band was lower and wider on harder concrete substrata. On porous brickwork walls *Enteromorpha intestinalis, E. prolifera* and *E. torta* replaced *Blidingia* at high water neap tide level.

The large brown algae *Fucus spiralis* and *F. vesiculosus* colonized lower levels of sea-walls and boulders on the foreshore. Fucoids were host to epiphytes such as *Blidingia marginata* and the small brown alga *Elachista fucicola.* Beneath the fucoid cover was a sporadic underflora of smaller red, brown and green algae (*Ceramium deslongchampsii, Pilayella littoralis, Rhizoclonium riparium* and *Ulva lactuca*). The xanthophyte *Vaucheria compacta* occasionally formed silt-binding patches on the sea-walls but was more often present on the foreshore.

At Greenhithe, 7 km further upriver, reappraisal revealed that the sloping boulder wall bore essentially the same flora as seen 15 years earlier with, additionally, minute amounts of *Urospora penicilliformis,* epiphytic *Elachista* and an underflora of *Polysiphonia urceolata* (*P. subtilissima* form).

The algal vegetation on sloping boulder sea-walls at Erith (nearer to London) supported three distinct bands of vegetation: at high water spring tide level and above, a band of halophytes (*Aster, Triglochin, Elymus*) overlapping the next band down, a bright green band of *Blidingia minima* and *B. marginata* and generally a dense band and canopy of *Fucus vesiculosus* extending almost to the foot of the sea-wall. There was no underflora apart from occasional small tufts of *Ulva lactuca* beneath *Fucus.* Velvety green patches of *Vaucheria* (cf. *compacta*) occurred among both *Blidingia* and *Fucus;* velvety patches of the red alga *Audouinella purpurea* were found at extreme high water spring tide level in the shade beneath piers and jetties.

Further upriver at Belvedere the zonation of algae on the river wall differed little from that at Erith except for the greens *Rhizoclonium riparium* and *Ulvaria oxysperma*, which also occurred on wood pilings. Reappraisal at Crossness revealed that the sloping boulder wall also differed little from that at Erith and bore a dense canopy of *Fucus vesiculosus*; at lower levels on the wall *Fucus* was replaced by mucilaginous silt-binding diatom growths. Further upriver at Tripcock Ness the river embankment comprised vertical sheet metal piling and this supported only a poor growth of algae, mainly *Blidingia marginata*. *Fucus* was sporadically present on foreshore boulders and on new concrete structures such as steps and slipways.

All sites between Tripcock Ness and the uppermost limit of the brackish-water estuary (about Barnes) were dominated by green algae, principally *Blidingia* and *Rhizoclonium*. Bellwater Gate at Woolwich, the site of detailed seasonal and zonal studies in the 1970s (Tittley, 1985), has been completely rebuilt, with the original brick walls largely replaced by sheet metal piling. Most algal growth occurred on concrete structures and was dominated by successive bands of *Blidingia* and *Rhizoclonium*; diatoms predominated at lower shore levels. Small, juvenile *Fucus* plants occurred amongst the *Blidingia*; sandwiched between the *Blidingia* and *Rhizoclonium* bands was a narrow one of *Ulvaria oxysperma*. A small area of remaining brick bore *Audouinella purpurea*. The overall species diversity was less than that recorded 15 years earlier. At the Thames Barrier the river wall vegetation was dominated by *Blidingia* and *Rhizoclonium*. Patches of *Vaucheria* were common while *Enteromorpha* spp. occurred only sporadically.

Floating piers provided two habitats: an upper wave-washed zone corresponding to a narrow intertidal area, and a lower submerged zone corresponding to the subtidal. The upper zone contained dense growths of mainly green algae (*Ulothrix, Enteromorpha* and *Ulva*), with plants of *Porphyra* and *Fucus* occasionally present; red algae such as *Ceramium deslongchampsii, Callithamnion roseum* and *Polysiphonia nigrescens* were common at shallow sublittoral levels. In 1986 the filamentous feather-like green alga *Bryopsis plumosa* was found growing on a floating pier at Tilbury (J.H. Price, personal communication).

The algae dominating saltmarsh communities remained the green algae *Blidingia* and *Enteromorpha* spp., the latter also a primary colonizer of bare mud. Pools and channels were often lined with *Vaucheria*.

4.3.2 Freshwater tidal reach

In this innermost part of the estuary the very few macroalgae present were found sporadically colonizing metal pilings, bank reinforcing walls or other hard surfaces and there was no well defined pattern of algal zonation. The only macroalgae observed here (*Blidingia marginata, B. minima, Rhizoclonium* spp. – possibly *R. hieroglyphicum* and *R. riparium*) were largely confined to foreshore areas above about mid-tide level. Commonly accompanying these green algae was *Vaucheria*, although it formed more extensive mats on stable mud banks – especially those found between Teddington and Richmond. In this stretch a greenish or blue-green powdery layer was common about high water spring level and consisted of the filamentous green alga *Klebsormidium flaccidum*, coccoidal cells (principally *Pseudendoclonium* spp.) and various blue-green algae.

Artificial settlement surfaces were quickly colonized by a wide range of freshwater algae that included species of *Ulothrix*, *Oedogonium*, *Protoderma* and *Stigeoclonium*. These green algae were confined to, or more frequent at, the partially tidal upriver site at Richmond lock than at two fully tidal sites downriver. The partially tidal freshwater stretch closely resembles the nontidal river in terms of species richness, composition and frequency of occurrence of individual taxa. Nonetheless, all sites in the freshwater tidal river were poorer in species than those examined upriver of the weir at Teddington. The downriver penetration of freshwater species has yet to be established precisely.

4.4 SEASONALITY

In the outer (marine) reaches of the estuary, seasonal changes are less obvious as the flora has a greater proportion of perennial species. A small but distinctive element comprised late winter–early spring ephemeral species such as *Dumontia contorta*, *Monostroma grevillei* and *Petalonia fascia*.

Distinct seasonal changes in algal growth were recorded by Tittley (1985) on a brick wall at Woolwich. These involved the die-back of an extensive *Enteromorpha* cover and its replacement initially by a mucilaginous diatom-dominated community, trapping and binding silt. Subsequently blue-green algae such as *Porphyrosiphon notarisii* and *Schizothrix calcicola*, together with the unicellular red alga *Porphyridium purpureum* and *Vaucheria compacta*, appeared on and in the mat of trapped silt; the intertwining filaments of the latter provided stability to the mat. The mat eventually sloughed off, revealing bare surface which was recolonized by *Enteromorpha*. These changes are summarized in the time-series ordination of dated transect recordings against which abundances of *Enteromorpha* and the reciprocal ordination of species have been overlain (Figure 4.2).

A similar sloughing away of mats of sediment and associated algae (principally blue-green algae) has been observed taking place in the freshwater tidal river during the summer and early autumn (John *et al.*, 1990). It is possible that the algal colonizers of the settlement plates are the same taxa that first develop on hard surfaces exposed after the sloughing-off of a smothering layer. Spring growths of diatoms on the wall studied by Tittley (1985) coincided with the peak of diatom numbers as recorded by

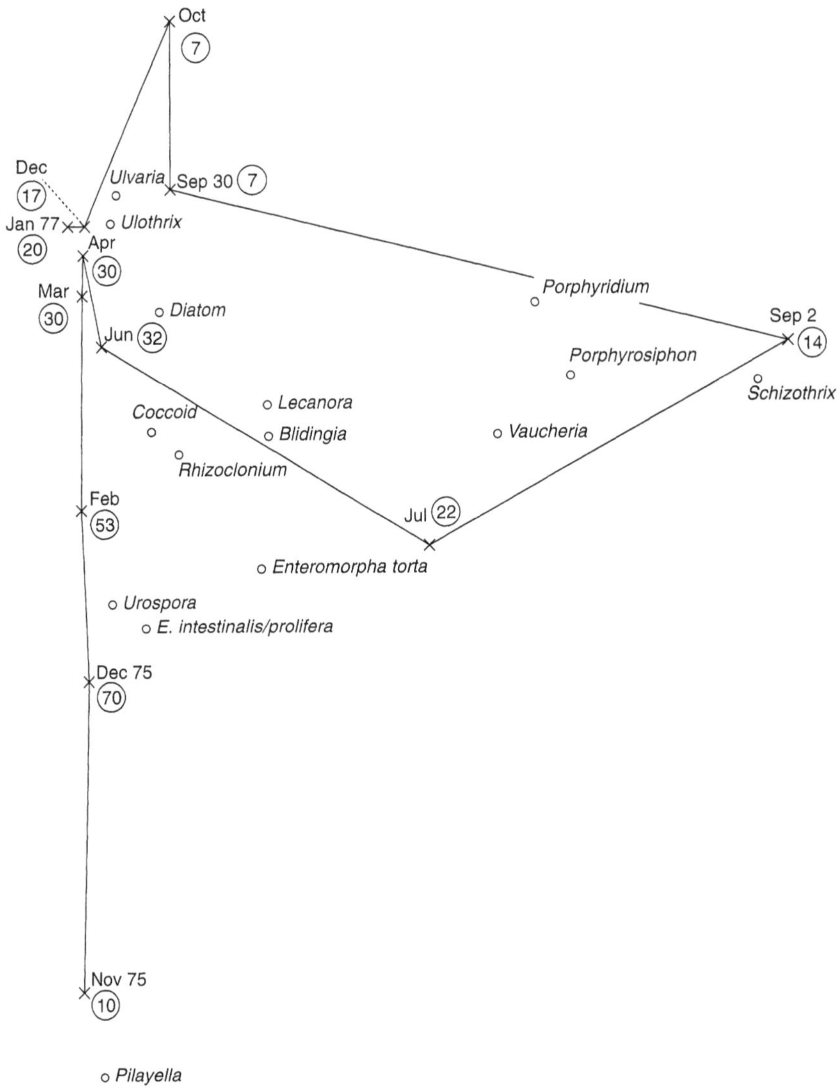

Figure 4.2 Time-series ordination (reciprocal averaging) of Woolwich transect data (from Tittley, 1985) with species ordinations overlain. x = dated observation; figures in circles = abundance of *Enteromorpha*.

Gameson and Johnson (1964) in the phytoplankton of the brackish reaches of the Thames.

Species richness on the settlement surfaces placed in the tidal freshwater reach tended to be lowest during the main flood period in February, peaked in May and showed an irregular decline over the remainder of the year. Seasonal changes were seen clearly in one of the most common macroalgae recorded in this stretch, *Vaucheria* spp. This alga grew on hard surfaces and was also the only macroscopic form to colonize relatively stable areas of mud. The dark green velvety covering of this xanthophyte often first became evident in March and April; extensive spongy mats developed by May and June, only to almost disappear for much of the remainder of the year. Its distinctive thick-walled spores were noticeable on settlement surfaces examined during the principal growth period (April–July).

Seasonal differences in the composition of the algae colonizing settlement surfaces placed in the freshwater tidal estuary relate to many factors, e.g. dilution of available propagules during flood periods, non-production of propagules, selectivity exhibited by surfaces and prevention of settlement due to water speed.

4.5 CHANGES IN OCCURRENCE AND DISTRIBUTION

A few algae, new to the estuary, have been recently recorded in the outer reaches, including the green algae *Bryopsis plumosa* and *Chaetomorpha capillaris* on floating piers and the filamentous brown alga *Ectocarpus siliculosus* as underflora on boulders at Greenhithe. The red algae *Polysiphonia nigrescens* and *Porphyra purpurea* now occur at Tilbury and Gravesend on floating piers, 20 km further upriver from their previously known limits where they grew on intertidal substrata. A form of the filamentous red alga *Polysiphonia urceolata* was also recorded further upriver at Greenhithe (this form should probably be attributed to *P. subtilissima*, a species not currently listed for the British Isles but which commonly grows in reduced salinities in warmer waters – taxonomic studies are in progress to resolve its status; C. Maggs, personal communication). The small brown alga *Elachista fucicola*, an obligate epiphyte on fucoids, now occurs at Greenhithe, which is 6.5 km upriver from its previously identified limit.

The most noticeable change in distribution has been the 8 km upriver migration (Belvedere to Woolwich) of *Fucus vesiculosus*. At Crossness (3 km west of Belvedere) a dense canopy of large fertile *Fucus* plants now covers the upper littoral zone on sea-walls where previously it was absent. Occasional small and fertile plants have also spread to steps and slipways further west at Tripcock Ness. At Woolwich the few, small juvenile plants on concrete steps represent its present upriver limit. The recent spread of *Fucus* to Woolwich represents a migration into an area of much lower salinity (7 g/l) than its previously known limit at Belvedere (16 g/l). *Ulva lactuca* has also migrated upriver, but not to the same extent as *Fucus*.

The shift in distributional limit of *Fucus* and *Ulva* in the middle reaches, and other algae in the outer estuary, may be due to a slow long-term response to the amelioration in river water quality, or to changes in salinity due to decreasing freshwater flow. A concurrent upriver migration of macrofauna species is described in Chapter 6.

New structures in the estuary have provided additional surfaces for algal colonization. The new port facilities at Dartford, including its floating piers, already support the pioneer colonizers *Enteromorpha* and *Ulva*; future monitoring will reveal whether or not this site is suitable for estuarine red and brown algae. The construction of river wall along almost the entire length of the estuary represents a major change in habitat for algal colonization. The saltmarsh communities dominated by green algae, formerly widespread, have now been replaced by rocky shore communities dominated by fucoids. The occurrence of *Ascophyllum* on sea-walls in the outer estuary is not unexpected and follows its widespread colonization of new habitats in the southern North Sea (Tittley 1985, 1986). Replacement of existing sea-walls with harder, impervious materials has had an impact on local ecology (see above).

Marsh ditches near Charlton and the Thames between Greenwich and Woolwich were the type locality for the xanthophyte *Vaucheria dichotoma*; specimens still exist from collections made in the eighteenth century (cf. Dillenius, 1742; Christensen, 1987). This species has not been found since in the estuary; it is probably now extinct due to the loss of fringing marsh habitat.

Several benthic macroalgae found in other estuaries remain absent from the Thames. A lack of suitable sites where freshwater directly enters marine conditions, lack of suitable substrata, excessive turbidity and pollution are factors that might account for the absence of the estuarine fucoid *Fucus ceranoides*. Another fucoid, *Pelvetia canaliculata*, absent in the outer estuary is of sporadic occurrence in the southern North Sea. The small green algae *Capsosiphon fulvescens* and *Percursaria percursa* and the crustose red alga *Hildenbrandia prototypus* are also absent despite the latter's occurrence in the nearby Medway estuary. The two common saltmarsh red algae *Bostrychia scorpioides* and *Catenella caespitosa* inhabit saltmarshes in the Medway but not the little remaining saltmarsh in the Thames. The absence of *Prasiola stipitata* is surprising, since the species is commonly reported in eutrophicated estuaries.

4.6 DISTRIBUTION ALONG THE ESTUARY

The distributions of benthic macroalgae at 36 sites along the Thames estuary are summarized in Table 4.1; this clearly shows a decrease in species number with increasing distance from the sea. The observations made by Tittley and Price (1977a) on the distribution of the principal algal groups still hold, namely that the Rhodophyta remain more diverse in and are mostly restricted to the fully marine outer estuary, that the Phaeophyta penetrate further into the middle reaches than the red algae, and that the majority of the Chlorophyta penetrate the brackish reaches, with a few extending into the tidal freshwater river. The latter is characterized by a distinct assemblage of freshwater species (John *et al.*, 1990, see above), principally green algae, including many microscopic forms. *Vaucheria* is also common in this reach, with the freshwater species *V. bursale* appearing to be confined to it.

Ordination of presence–absence data using DECORANA does not show widely separate clusters but a strong linear spread with a small separation into groups of sites (Figure 4.3) along the first axis; this axis represents a decreasing salinity gradient and correlates with a reduction in species number. The spread of points exceeds four standard deviations and thus indicates significant floristic differences between the extremes. The ordination shows marine sites merging into a tighter central cluster of sites in the river from Northfleet to Chelsea. A small group of sites having high values on the first axis are mostly those between central London and Putney. The highest values on the first axis are for sites in the tidal freshwater river.

The association of Purfleet (site 14) with inner Thames estuary sites is due to a species-poor flora common to sites throughout the estuary. Conversely the association of Chelsea (28) with downriver sites is due to its slightly higher species-richness. The association of Dagenham (21) with outer estuary sites reflects the occurrence of *Fucus spiralis*, a brown alga otherwise more common in the outer estuary. The spread along the second

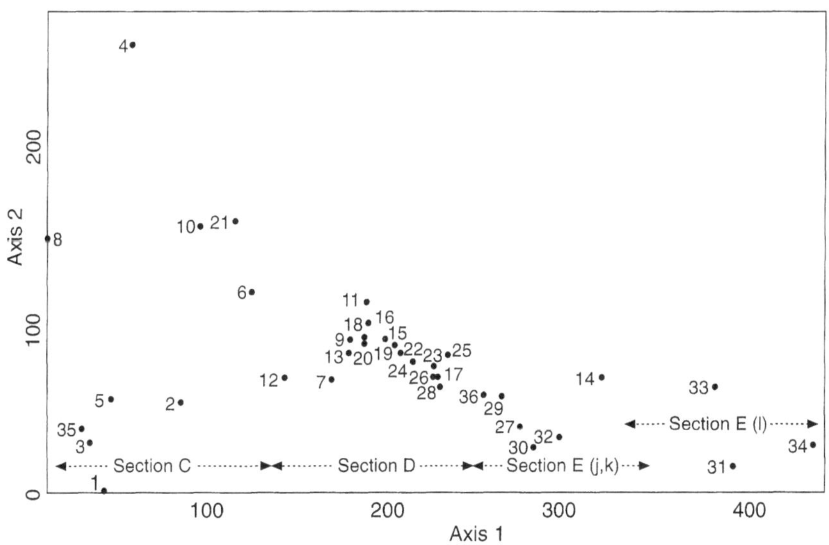

Figure 4.3 Ordination (DECORANA) of 1992 Thames estuary marine algal data. Floristic sections D–E and subdivisions J–L indicated (see Figure 4.4).

Table 4.1 Algal species recorded at sites along the Thames estuary

Species Site	1	2	3	4	5	6	35	7	8	9	10	11	12	13	14	15	16	17	18	19	20	21	22	23	24	25	36	26	27	28	29	30	31	32	33	34
Audouinella daviesii	#																																			
Eugomontia sacculata	#																																			
Callithamnion hookeri	#																																			
Dumontia contorta	#	#																																		
Porphyra purpurea	#		#				#																													
Polysiphonia nigrescens	#		#				#		#																											
Erythrotrichia carnea			#																																	
Chondrus crispus	#	#	#	#																																
Petalonia fascia				#																																
Monostroma grevillei				#																																
Ralfsia clavata	#	#		#	#																															
Audouinella floridula	#				#																															
Gelidium pusillum	#	#	#		#																															
Audouinella sp.					#																															
Polysiphonia urceolata					#																															
Bryopsis hypnoides									#																											
Chaetomorpha capillaris									#																											
Ceramium deslongchampsii	#	#	#		#		#	#	#				#																							
Ectocarpus siliculosus													#																							
Callithamnion roseum							#	#	#																											
Cladophora fracta									#																											
Elachista fucicola	#					#		#		#					#																					
Bangia atropurpurea													#																							
Ulothrix flacca												#														#										
Cladophora sericea	#	#												#																						
Ascophyllum nodosum	#		#		#																															
Enteromorpha flexuosa														#		#		#																		
Ulva lactuca	#		#	#	#	#	#	#	#	#		#	#	#		#	#	#	#			#	#													
Fucus spiralis	#		#	#	#	#		#	#	#	#	#	#	#		#	#					#														
Fucus vesiculosus	#	#	#	#	#	#	#	#		#	#	#	#					#		#	#	#	#	#		#	#									
Ulvaria oxysperma								#		#													#			#	#									
Pilayella littoralis	#			#	#	#		#	#	#		#	#	#		#	#	#	#		#		#		#	#										
Audouinella purpurea					#	#		#								#	#		#	#	#		#		#	#										
Enteromorpha torta						#		#						#			#	#	#		#		#		#	#	#	#								

Species Site	1	2	3	4	5	6	35	7	8	9	10	11	12	13	14	15	16	17	18	19	20	21	22	23	24	25	36	26	27	28	29	30	31	32	33	34
Porphyridium purpureum																										#										
Enteromorpha prolifera		#	#		#	#	#	#	#	#	#	#	#	#		#	#	#	#		#	#	#	#	#	#	#	#		#	#					
Enteromorpha intestinalis	#	#		#	#	#	#	#	#	#	#	#	#	#		#	#		#	#	#	#	#		#	#	#	#		#						
Ectocarpus sp.																											#									
Urospora penicilliformis	#							#		#		#	#			#	#	#			#		#		#	#						#	#			
Vaucheria compacta						#		#		#		#		#		#	#	#	#				#	#	#	#	#		#	#				#		
Blidingia minima	#	#	#					#		#		#	#	#		#	#	#	#	#	#		#		#	#	#	#	#	#		#				
Blidingia marginata	#	#	#	#				#		#		#	#	#	#	#	#	#	#	#	#	#	#	#	#	#	#	#	#	#	#	#			#	
Rhizoclonium riparium	#	#				#		#		#		#	#	#	#	#	#	#	#	#	#		#		#	#	#	#	#	#	#	#	#	#	#	#
Cladophora fw. sp.																																	#	#	#	#
Average salinity (ppt)	31	31	31	31	30	25	24	24	24	24	24	22	19	19	19	19	18	18	18	16	16	16	16	15	13	7		5	4	3	3	2	2	2	0	0
Salinity range (ppt)	(25–32)							(14–2€)													(2–16)								(0–3)							
Distance from Grain (km)						24						32							44							54					74					91

axis of outer estuarine sites (Grain to Gravesend) reflects their floristic diversity with considerable local variation in species of Chlorophyta, Phaeophyta and Rhodophyta, the latter being a prominent component of the flora. In contrast, sites in the tidal freshwater river are characterized by a species-poor green algal flora and by a distinct assemblage of freshwater microspecies.

DECORANA ordinations of the 1977 and 1992 data sets showed a similar overall structure; a small number of sites were repositioned in the 1992 analysis due to new species records. The similarity in the ordinations indicates the consistent broad distribution patterns of principal algal groups in the estuary. An ordination of surface sediment diatom samples from sites between Teddington and Shoeburyness (Juggins, 1992, Figure 4.5) also showed a clear linear distribution which strongly related to salinity levels. Ordination of macroinvertebrate sampling sites throughout the Thames estuary (Chapter 6) likewise shows sites principally distributed along one axis and, as with the algae, faunistic sections can be recognized.

An analysis using TWINSPAN classified the sites into two main groups (Figure 4.4): group A comprised outer estuary sites to Greenwich, and group B inner estuary sites from central London to Barnes. The former group is further subdivided into species-richer mainly outer estuarine sites (group C), and species-poorer middle reach sites from Gravesend to London (group D). Further dichotomies subdivided the main sections as follows:

- floating pier sites at Tilbury and Gravesend and the Tilbury and Dagenham Dock sites (group f, characterized in part by their sublittoral species assemblage) from other outer estuarine sites (group g);
- four outer middle reach sites (group h) from the remaining middle reach sites (group i);
- the tidal freshwater river sites (group l) from those situated between central London and Putney (group k).

Analysis using PAUP produced ecocladograms which gave similar results to the ordination analyses but different floristic boundaries could be inferred. That a large number of equally parsimonious ecocladograms were produced suggests variability in floristic relationships between sites; however, comparison of ecocladograms reveals a consistent overall structure except for the poor resolution of floristically similar, species-poor inner sites. Branches representing species-rich sites are positioned apically in the ecocladograms while those representing species-poorer sites are basal.

Four floristic sections can be inferred from the ecocladograms (an example is presented as Figure 4.5):

- species-rich marine and floating pier sites (corresponding to section C in the TWINSPAN analysis), a diverse group indicated by wide spacing between dichotomies (Figure 4.5);

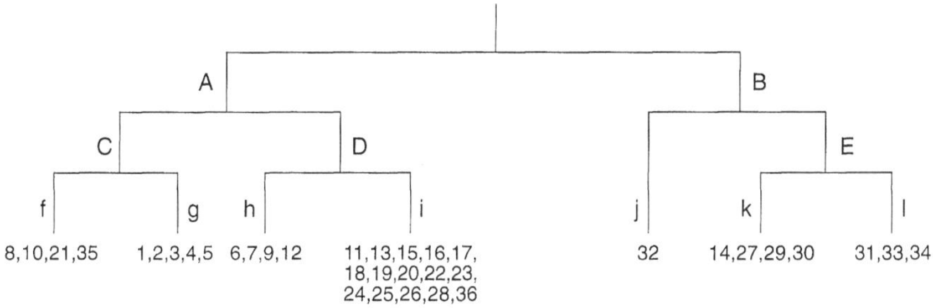

Figure 4.4 TWINSPAN dendrogram of 1992 Thames estuary marine algal data. **A–E** = main floristic sections; **f–l** = subdivisions of sections.

- estuarine sites with a lower species diversity (corresponding to part of section D in the TWINSPAN analysis);
- brackish middle-reach sites (comprising the remainder of D);
- inner estuary sites (corresponding to section E).

The PAUP results, in contrast to those of DECORANA and TWINSPAN, suggest a possible boundary between sites 18 and 20 (Erith and Belvedere) which are close to where Tittley and Price (1977a) placed a boundary; another is suggested for the Greenwich reach. Sectional boundaries were placed where internodal distances were larger than one.

Tittley and Price (1977b) attempted to correlate algal distribution with salinity and divided the estuary into four sections, each with a 'characteristic algal flora'. These were not strongly delineated and merged into one another, but appeared to relate to divisions defined in the 'Venice System for the Classification of brackish waters' (Den Hartog, 1960). Tittley and Price (1977a) used the upriver limit of *Fucus* as a criterion for delimiting the inner limit of the estuarine section; its upriver migration means that the boundary would need to be repositioned. Price (1982) identified an early stage in this migration and also commented on the boundary.

DECORANA analysis indicates clearly a gradient of floristic change from marine to freshwater conditions, whereas from TWINSPAN and PAUP floristically defined sections can be inferred (which also correspond to the ordination), thus refuting the null hypothesis that no such sections exist in the Thames estuary.

This reappraisal suggests that the best-defined and most strongly contrasting floras are restricted to the freshwater and fully marine reaches; this is clearly indicated in all analyses. There are no species restricted to the middle brackish reaches: they are defined more by the absence of species – the macrophytic flora of the inner estuary is an impoverished marine flora. All analyses suggest a marine section corresponding to that of Tittley and Price (1977a). Ecocladistic analysis suggests an estuarine section to

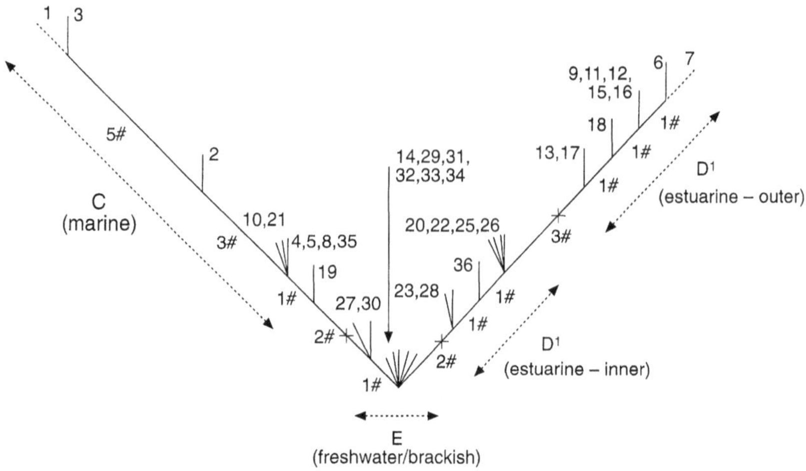

Figure 4.5 Scaled ecocladogram showing internodal distance (#); X = divisions (where internodal distance exceeds one unit*) into groups of sites (estuarine sections); **C**, **D**, **E** = sectional divisions indicated by TWINSPAN analysis. [*The large internodal distances between the marine sites indicate considerable inter-site variation (see also DECORANA ordination); sites are agglomerated into a single marine section for convenience.]

Erith/Belvedere which also corresponds to that of Tittley and Price (1977a), while the ordination analyses suggest a longer section to the Greenwich reach. All three analyses suggest that a boundary area between the estuarine and brackish/freshwater sections lies in the Greenwich area and the occurrence of *Fucus* appears not to be the delimiting factor.

TWINSPAN suggests further subdivisions of the three sections. The boundary placed in the Barnes area by Tittley and Price (1977a), and identified by TWINSPAN as an inner estuary subdivision, is probably more significant due to the presence of freshwater microphytes in the freshwater tidal river (data for freshwater algae were not included in the analyses).

4.7 FUTURE STUDIES

The potential distributional limits of both freshwater and marine benthic algae in the tidal estuary remain to be established. To this end more precise information on the distribution of microphytic species or microscopic stages needs to be obtained by the use of artificial settlement plates. Care needs to be exercised in positioning such surfaces, as any placed on gently sloping embankments below mid-tide level would soon become covered with a smothering layer of fine sediment. To use the distribution of benthic micro- and macroscopic algae to monitor longer-term changes in water quality, it would be necessary to eliminate as far as possible effects caused by differences in substratum type and periods of emersion and immersion. This might be achieved by standardizing the surface used and fixing these either at the same level in the intertidal or on floating structure at a constant depth below the water.

The macroalgal flora needs to be monitored routinely and general observations should be augmented with recording at permanently established sites throughout the estuary. This information would make it possible to measure floristic change or stasis and tolerate any such change to variations in the physical and chemical environment.

The investigation by Juggins (1992) has demonstrated that there is a diverse community of diatoms associated with sediments in the Thames estuary. Information is required on other algal groups inhabiting sediments to establish whether the total living assemblage of algae forming the epipsammon and epipelon reflect surface water quality in the estuary. In future more attention should be directed towards the phytoplankton which, like the sediment-inhabiting forms, have been much neglected.

4.8 CONCLUSIONS

Very little natural embankment remains in the Thames estuary. A marshy soft-sediment environment (the algal component formerly characterized by Chlorophyta and other saltmarsh species) has been converted to a rocky shore environment where fucoids are now commonly present in the middle and outer reaches. Artificial structures now provide a wide range of intertidal and subtidal habitats for algal colonization.

The estuary is a dynamic ecosystem exhibiting both short-term (seasonal) and longer-term (migrational) changes in benthic algal populations.

Artificially placed surfaces provide a valuable means of recording the otherwise overlooked microphytic component of the algal flora; the extension of this technique to the tidal estuary will help to define more precisely the upriver and downriver limits of species distributions.

The diversity of benthic algae declines from the outer to the inner estuary only to increase again in the freshwater reaches where truly freshwater algae are present. A gradual change occurs in algal flora from outer marine conditions to inner freshwater conditions, but it is the extremes that are floristically most distinct. Floristically defined sections can still be inferred but must be viewed with caution; those in the middle reaches are less precise and boundaries may be mobile. The freshwater tidal reach is characterized by the dominance of green algae; brown and red algae are largely restricted to the marine reaches, with the exception of a few species tolerant of lower salinities.

CHAPTER 5

The zooplankton communities of the Thames estuary

Chris Gordon, Anthony Bark and Roland Bailey

5.1 INTRODUCTION

Zooplankton comprises those animals which live suspended in the water column with limited powers of movement such that they drift with water currents. Some spend their entire lives in suspension (holozooplankton) whereas others only inhabit the water column for part of their life cycle (merozooplankton). Estuarine zooplankton has an autochthonous component, which develops and remains in the estuary, and an allochthonous component comprising freshwater organisms washed downstream by the river and marine organisms that have penetrated the estuary from the coastal waters. The presence of allochthonous zooplankton typically results in estuaries possessing high species diversity in the upper and outer ends and having a species-depauperate fauna in the middle reaches, a pattern similar to that recorded for benthic invertebrates (Chapter 6).

The Thames estuary is large, unstratified, canalized, turbid and macrotidal (tidal range greater than 4.0 m), has a large tidal excursion (*c*. 15 km) and rapid current speeds (*c*. 3.5–7.5 km/hour; HMSO, 1964; Chapter 3) which presents peculiar problems both to the zooplankton and to those who wish to study them. In this chapter, previous work carried out on zooplankton in the Thames estuary is reviewed, and the results of a 27-month investigation of zooplankton distribution in the estuary are presented and discussed. The taxonomic composition and the spatial and temporal distributions of zooplankton in the Thames are compared with those in other large estuaries.

A Rehabilitated Estuarine Ecosystem. Edited by Martin J. Attrill.
Published in 1998 by Kluwer Academic Publishers, London. ISBN 0 412 49680 1.

The river Thames, and its estuary, is one of the best documented rivers in the world with studies covering all aspects of physical, chemical and biological processes (IMER, 1984a,b) but, at the same time, little information is available on the abundance and distribution of zooplankton in the tidal Thames. The earliest work was that of Wells (1938), based on samples taken from the end of Southend pier. Information is presented in his semi-quantitative survey of the outer estuary, on circulation patterns, the zooplankton species present and their temporal distributions with season and tide. Despite the limitations of shore-based sampling Wells (1938) gave a good description of the plankton in the outer estuary. El-Maghraby (1956) studied the plankton in the estuary between Whitstable and the Isle of Sheppey, focusing chiefly on diatoms but with some valuable observations on the diversity of the microzooplankton. Lumkin (1971) investigated the relationship between pollution and plankton from Barking Reach to Barrow Deep. Using a pump sampler on board sewage sludge vessels, he was able to sample a 160 km stretch of estuary in a salinity range from 10% to 34%. He noted the importance of calanoid copepods in the estuary and devised a classification scheme for the Thames estuary based on zooplankton communities.

5.2 THE CURRENT STUDY ON THE ZOOPLANKTON OF THE THAMES ESTUARY

The current study looks at the meso-zooplanktonic fauna (size: 0.1–10.0 mm; Burkill, 1983) at nine sites along the tideway from Kew to Tilbury. No work on the zooplankton of this stretch of the estuary has been published before. Use has been made of the Environment Agency (EA) half-tide correction model (IMER, 1984a; Chapter 2) to normalize the samples taken at a particular site to EA water zones. Figure 5.1 gives the half-tide position of zones in which the sampling sites were grouped. The half-tide correction factors enable the site of sampling to be standardized in respect to the water mass at any given time. The salinity of the water sampled in the study ranged from 0.4% to 24.0%.

Samples were taken with a 0.335 mm mesh net fitted with a flow meter, towed by boat in the top one metre of the water column. Three samples were taken at each site, usually off one bank, except in the area around London bridge, where alternate banks had to be used due to heavy boat traffic and Port of London Authority navigation rules. Tows lasted between four and five minutes and about 10 m^3 of water were filtered per sample. Pump samples were also taken over a 24-hour period to assess the significance of vertical migration. A 50 mm high-volume centrifugal pump was used for this and samples were taken from the top and the bottom of the water column each hour.

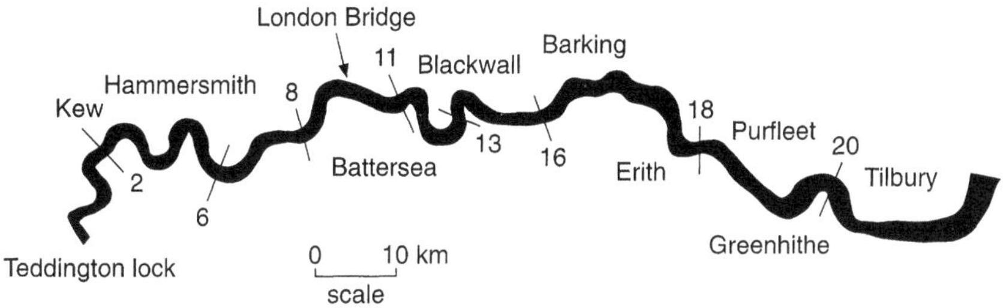

Figure 5.1 Map of the Thames estuary showing sampling zones at half tide based on the EA tide correction model. Numbers refer to limits of sampling stretches in the 1990–1992 survey.

5.3 COMPOSITION OF THE ZOOPLANKTON

A full list of species recorded as plankton in the Thames estuary is given in Appendix B, which is a synthesis of our own studies and those of Wells (1938), El-Maghraby (1956) and Lumkin (1971). In all, more than 150 planktonic taxa have been identified over the past 50 or so years. A list of the major species recorded in our study is presented in Table 5.1, showing the meroplanktonic and holoplanktonic components. As a result of the large amounts of suspended material in the tideway which prevent the efficient use of small mesh sizes in net-based sampling gear, small species are underrepresented in the samples.

5.3.1 Protozoa, Cnidaria, Ctenophora and Rotifera

Apart from diversity data, very little is known about the protozoan communities of the Thames estuary. Wells (1938) noted the presence of foraminiferans in all samples, with *Polystomella striato-punctata* being the most frequent, followed by *Miliolina* sp., *Biloculina* sp., *Nonionina* sp., *Planorbulina mediterranensis*, *Lagena striata* and *Textularia variabilis*, in that order. Dinoflagellates, mainly species of *Ceratium* and *Peridinium*, were also present. El-Maghraby (1956) found large quantities of *Tintinnopsis* sp. at Whitstable (over 1000/m^3), particularly during the summer months.

Cnidarians were present in variable numbers during the period 1989–1992; *Aurelia aurita* was the most common species and in early summer large numbers were found on the fish screens of the West Thurrock power station (Attrill and Thomas, 1996). Wells (1938) found significant quantities of *Rathkea octopunctata* and *Phialidium hemisphericum* in the Southend area. Only one species of ctenophore, *Pleurobrachia pileus*, has been recorded from the estuary by all workers on the Thames, with maximum numbers in summer samples (Attrill and

Table 5.1 Common zooplankton found in the Thames Estuary Study, 1990–1992

	Position in estuary					
	Zone 2–7 Upper		*Zone 8–17 Middle*		*Zone 18–24 Lower*	
Species	*Adults*	*Larvae*	*Adults*	*Larvae*	*Adults*	*Larvae*
Holoplankton						
Aurelia aurita				×		×
Pleurobrachia pileus			×	×	×	×
Brachionus quadridentatus	×					
Keratella quadrata	×					
Bosmina longirostris	×					
Daphnia longispina	×	×				
Pleuroxis uniciatus	×					
Acartia bifilosa	×		×		×	
Centropages hamatus			×	×	×	×
Diaptomus gracilis	×	×				
Eurytemora affinis	×	×	×	×	×	×
Temora longicornis					×	×
Acanthocyclops viridus			×	×	×	×
Cyclops strenuus	×		×		×	
Cyclops vernalis	×		×		×	
Cyclops vicinus	×		×		×	
Eucyclops agilis			×		×	
Eucyclops serulatus			×		×	
Macrocyclops albidus			×		×	
Meroplankton						
Gastrosaccus spinifer					×	×
Mesopodopis slabberi						×
Neomysis integer		×	×	×	×	×
Corophium volutator			×		×	×
Gammarus pulex	×	×				
Gammarus zaddachi	×	×	×	×	×	×
Crangon crangon				×		×

Thomas, 1996). *Beroe cucumis*, reported by Lumkin (1971), occurs sporadically in early autumn.

Most rotifers found in the estuary are washed into its upper reaches, especially when large-scale river flushing follows a period of low discharge. In summer, with low flows, the river above Teddington weir (and to some extent above Richmond half-lock) may become virtually stagnant, allowing rapid development of rotifers in the lake-like conditions (Bowen and Pinless, 1977). No rotifers have been reported from the middle reaches; however, El-Maghraby (1956) and Lumkin (1971) found *Brachionus* sp., *Keratella* sp. and *Synchaeta* sp. at densities greater than $3000/m^3$ in the outer estuary in early spring. These may have been washed into the Thames estuary from some of the smaller rivers that enter its lower reaches.

5.3.2 Cladocera

The cladoceran assemblage was dominated by daphniids. They occurred in the tideway, very much like the rotifers, as survivors from the freshwater inflow into the Thames tideway from the non-tidal river. *Daphnia longispina* and *Bosmina longirostris* were the most common species. In spring, at times of high river flow over Teddington weir, *D. longispina* was found as far down as London bridge. The brackish water species *Podon intermedius* and *Sarsiella zostericola* were found by Lumkin (1971) in the area 70 km below London Bridge.

5.3.3 Copepoda

Calanoids were the most abundant copepods in the tideway, with three groups present, conforming in part to the generalized classification of plankton according to salinity in estuaries (Jefferies, 1967; Collins and Williams, 1981, 1982); that is:

- wholly estuarine – characterized by *Eurytemora affinis;*
- estuarine and marine – characterized by *Acartia bifilosa;*
- euryhaline marine – characterized by *Centropages hamatus.*

We did not take samples in the stenohaline marine zone but the work of Wells (1938) and Lumkin (1971) indicate that this community would be represented by the neritic *Pseudocalanus elongatus. Eurytemora* dominates the meso-zooplanktonic fauna of the estuary. The species present was referred by some of the earlier workers as *E. hirundoides* (Nordquist), but Busch and Brenning (1992) have clarified the nomenclature of the species and it is now correctly placed as *E. affinis* (Poppe). The freshwater calanoid *Diaptomus gracilis* is carried in river water into the upper reaches of the tideway in late spring. Small numbers of cyclopoid and harpacticoid copepods were found in all areas of the tideway. The most common cyclopoids found in our study were *Cyclops vicinus* and *Macrocyclops albidus.* Only five harpacticoids were found in our survey; these were all from different species, and occurred in the more marine environment. Wells (1938) and El-Maghraby (1956) found 24 species of harpacticoid copepods, the most common being *Alteutha* sp.

5.3.4 Mysidacea

In our study, mysids were found up the estuary as far as Kew (0.4%), although as a rule they were generally restricted to a narrow stretch further down the estuary in the Rainham area, where they occurred in appreciable numbers. *Neomysis integer* was most frequent, followed by *Mesopodopis slabberi. Gastrosaccus* sp. also appeared sporadically during late summer. El-Maghraby (1956) found that *Mesopodopis slabberi* was the most common species in the Whitstable area.

5.3.5 Amphipoda

After calanoids, gammarids were the next most abundant zooplankton group in our study, with *Gammarus zaddachi* showing the widest distribution. In summer, though meroplanktonic, it was the dominant component of zooplankton biomass in the upper reaches of the estuary, with individuals ranging in size from 0.1 mm up to 1.5 cm. *Gammarus pulex* could be found occasionally in early spring around Kew. *Corophium volutator*, though never found in large quantities, was present in a number of samples taken between London Bridge and Tilbury.

5.4 SPATIAL AND TEMPORAL DISTRIBUTIONS OF THE MAJOR GROUPS

5.4.1 *Eurytemora affinis*

The most abundant calanoid copepod, *Eurytemora affinis* showed a consistent seasonal pattern, (Figure 5.2), with peaks of abundance in late autumn and in early spring. A bimodal spatial distribution also appeared in spring 1990, with abundance peaks in zones 11–12 as well as 16–17, shifting less clearly in spring 1991 to peaks in zone 8–10 and 13–15.

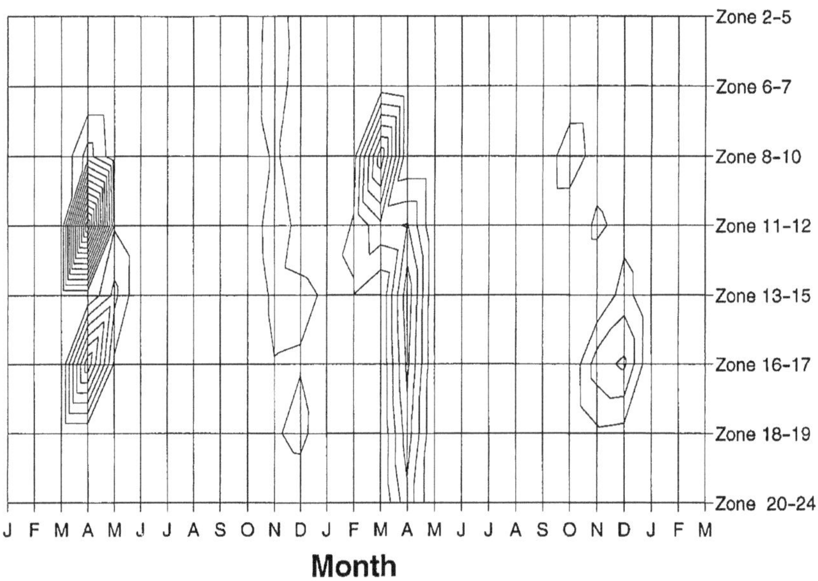

Figure 5.2 Spatial and temporal distribution of *Eurytemora affinis* between January 1990 and March 1992: isogram interval at 1000 individuals/m^3.

Examination of the relative abundance of the five major calanoids (*Acartia*, *Centropages*, *Diaptomus*, *Eurytemora* and *Temora*) in winter and summer (Figure 5.3) emphasizes the overall predominance of *E. affinis*. In summer, in zone 20–24 its numbers fell to less than one individual/m^3 and *Acartia bifilosa* assumed the dominant role.

5.4.2 *Gammarus zaddachi*

Numbers of *Gammarus zaddachi* peaked in late spring and early summer (Figure 5.4), with the most persistent tideway populations occurring above the Isle of Dogs. The zooplankton between Kew and Battersea was often completely dominated by this species, and in the area between Hammersmith and Battersea, one could often see adults trapped in pools at low tide. Large quantities of *Gammarus zaddachi* can also be found under stones in the intertidal regions in this area (Chapters 6, 8) and dominate the subtidal benthic community (Chapter 6).

5.4.3 *Neomysis integer*

More frequently found below London bridge than above it, *Neomysis integer* was generally concentrated in the middle reaches of the estuary,

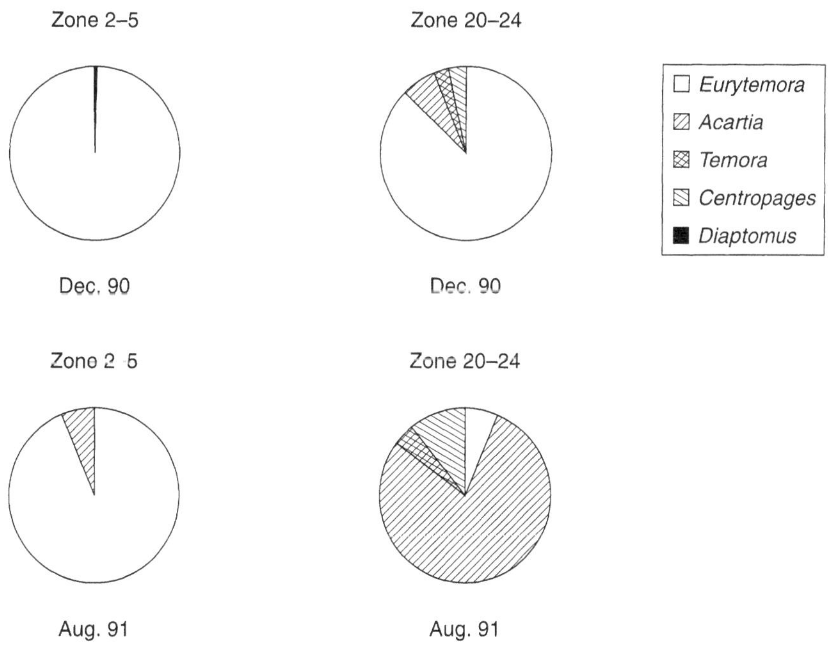

Figure 5.3 Comparison of the relative abundance of calanoid copepods in Zones 2–5 and 20–24 during winter and summer months.

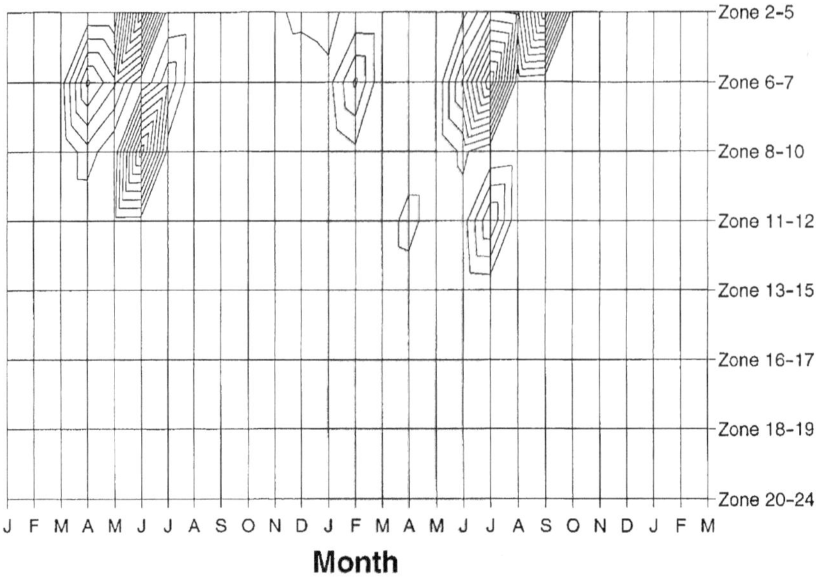

Figure 5.4 Spatial and temporal distribution of *Gammarus zaddachi* between January 1990 and March 1992: isogram interval at 30 individuals/m^3.

with the animals spreading further up and down the estuary in summer (Figure 5.5). Peak numbers of over 100 individuals/m^3 were recorded in the area between Beckton and Crossness, but it was more usual to find just one or two individuals/m^3 for most of the year. Very few ovigerous females were found and samples mainly comprised juveniles up to 12 mm in length.

5.4.4 Cyclopoids

Very low numbers of cyclopoids were found in the study area (Figure 5.6). In 1990 numbers were always less than five individuals/m^3. In spring 1991, *Cyclops vicinus* was found in zones 2–7, with sporadic occurrences further downstream. The relatively large mesh size of the sampling gear probably led to an underrepresentation of cyclopoids in the 1990–1992 field survey.

5.5 CHANGES IN DOMINANCE OF MAJOR SPECIES

A summary of the differences between the summer and winter fauna in the stretch of estuary between Kew and Tilbury is given in Figure 5.7. The transition from freshwater species such as *Diaptomus* sp. and *Daphnia* sp.

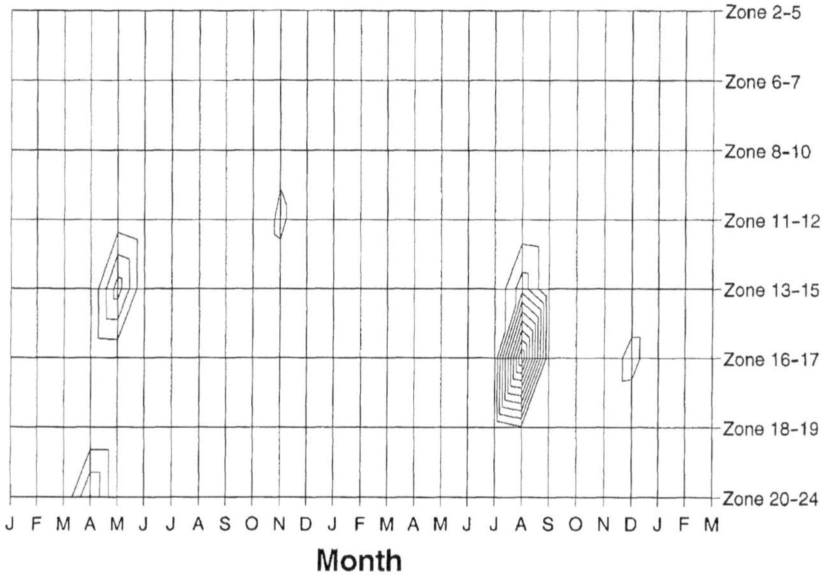

Figure 5.5 Spatial and temporal distribution of *Neomysis integer* between January 1990 and March 1992: isogram interval at 10 individuals/m^3.

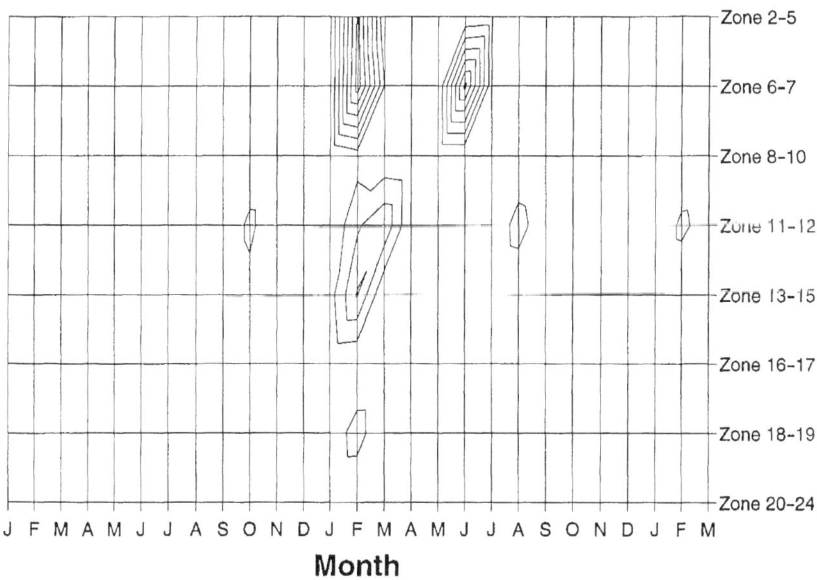

Figure 5.6 Spatial and temporal distribution of cyclopoid copepods between January 1990 and March 1992: isogram interval at 5 individuals/m^3.

Summer fauna

Occasional	*Diaptomus*	*Neomysis*	*Neomysis*	*Gammarus*	*Corophium*	Cyclopoids	Ctenophores	*Eurytemora*	**Occasional**
	Eurytemora	*Acartia*	*Gammarus*	*Neomysis*	*Neomysis*	*Neomysis*	*Eurytemora*	*Temora*	
	Daphnia	*Eurytemora*	*Acartia*	*Eurytemora*	*Eurytemora*	*Eurytemora*	*Centropages*	*Centropages*	
Dominant	*Gammarus*	*Gammarus*	*Eurytemora*	*Acartia*	*Acartia*	*Acartia*	*Acartia*	*Acartia*	**Dominant**
EA zone	**2–5**	**6–7**	**8–10**	**11–12**	**13–15**	**16–17**	**18–19**	**20–24**	
Dominant	*Eurytemora*	*Eurytemora*	*Eurytemora*	*Eurytemora*	*Eurytemora*	*Eurytemora*	*Eurytemora*	*Eurytemora*	**Dominant**
	Gammarus	*Gammarus*	*Gammarus*	*Gammarus*	*Neomysis*	*Gammarus*	*Acartia*	*Neomysis*	
	Diaptomus	*Neomysis*	*Neomysis*	*Neomysis*	*Gammarus*	*Neomysis*	*Centropages*	*Acartia*	
Occasional	*Daphnia*	Cyclopoids	Cyclopoids	Cyclopoids	Cyclopoids	*Acartia*	*Neomysis*	*Temora*	**Occasional**

Winter fauna

Figure 5.7 Summer and winter zooplankton of the Thames estuary from Kew to Tilbury (1990–1992).

to the euryhaline, marine species like *Centropages* sp. and *Temora* sp. is quite clear in summer. In winter, *Eurytemora affinis* dominates all zones of the estuary.

5.6 VERTICAL AND LATERAL DISTRIBUTIONS

Very limited work has been carried out on vertical migration of zooplankton in the Thames. Lumkin (1971) took hourly samples from one fixed depth of 4.2 m over a 24-hour period on one of his regular 380 km boat runs. He interpreted changes in abundance of *Eurytemora affinis* as movement up and down the water column in response to a diurnal light rhythm. In the present study, results from hourly samples taken at Waterloo pier at the surface and from the bottom of the water column over a 24-hour period in November 1992 (Figure 5.8) indicated that there was little difference in the vertical distribution of organisms at this site with time of day.

Due to the tidal excursion, sampling from the shore at one point over a 24-hour period effectively samples several different water masses. Table 5.2 links mean abundance of *E. affinis* and *G. zaddachi* from the surface and bottom of the water column with EA zones both for day-time and night-time samples. It can be seen that *Gammarus* is evenly distributed both vertically and between zones during the sampling period, while *Eurytemora* changes in abundance by two orders of magnitude from zone 8 to zone 12. It is possible that the change in numbers observed by Lumkin (1971) was due to sampling through a cloud of *Eurytemora* rather than actual vertical movement on the part of the plankton. By contrast, similar pump-sampling techniques on other estuaries have shown pronounced differences between zooplankton numbers in surface and bottom waters. This could be linked to time of day, with higher numbers at the surface at night (Fulton, 1984); other workers – for example, Vuorinen (1987) – suggest that this migration is linked to an escape response from predation. As a consequence of its large tidal range (4.6 m at Tilbury and 5.0 m at Kew) in our study area and the concomitant high current speeds of 1–2 m/s, the Thames is very turbulent (HMSO, 1964). Thus, apart from very short periods at slack tide, the water column is in a continual state of flux and in such situations the plankton are very much at the mercy of the currents (Barlow, 1955).

Differences in the abundance of zooplankton have been observed in the Thames at Blackwall between the north and south banks (Figure 5.9) but this is more likely to be due to differential rates of current speeds round bends and Coriolis forces in the channel (Preddy, 1954) than active migration. Heckman (1986) described the anadromous migration of *Eurytemora affinis* from the main channel to shallow canals and the sublittoral plant cover in the Elbe estuary. This was not observed in the Thames, which has

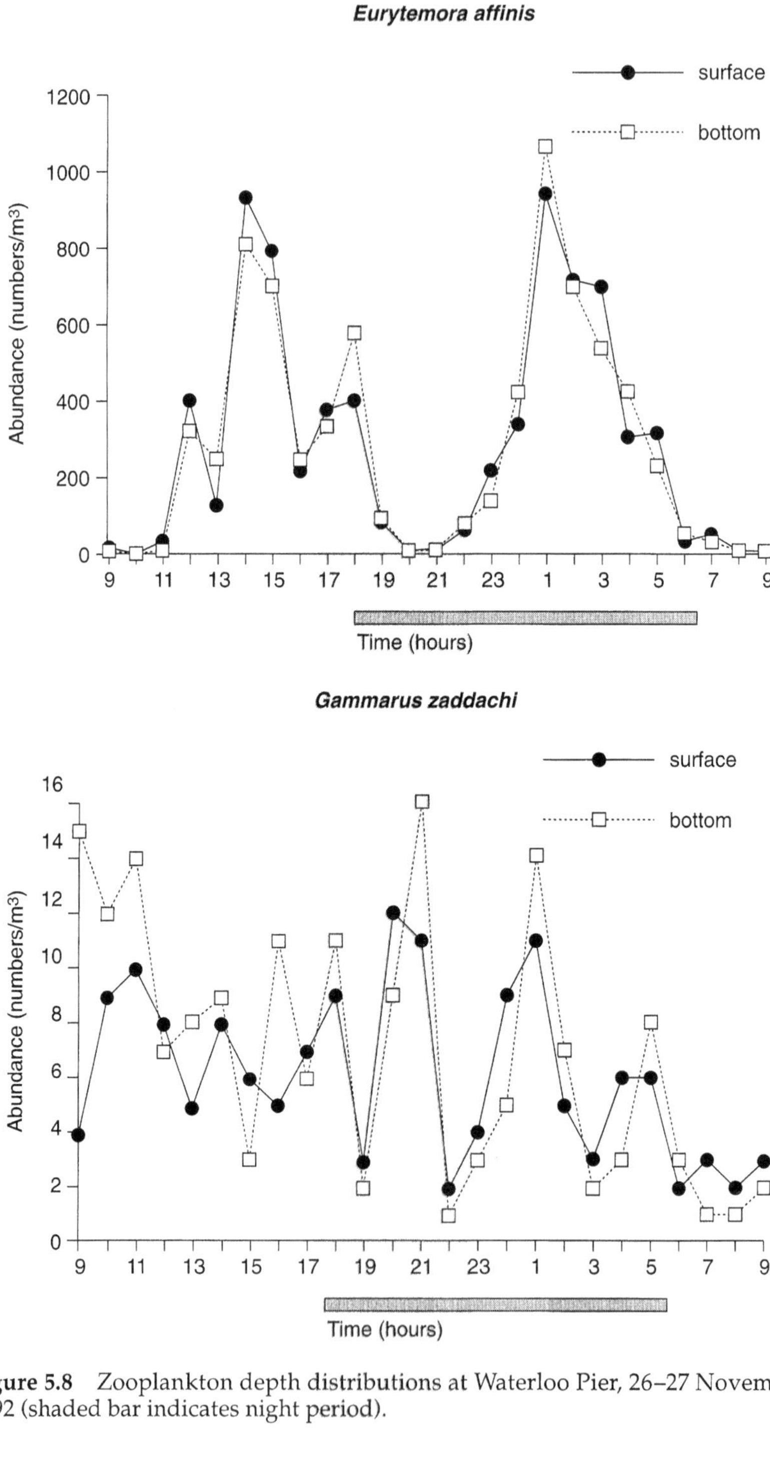

Figure 5.8 Zooplankton depth distributions at Waterloo Pier, 26–27 November 1992 (shaded bar indicates night period).

Table 5.2 Day and night-time depth distributions of *Eurytemora affinis* and *Gammarus zaddachi* sampled with a plankton pump at Waterloo Pier, 25–26 November 1992

EA zonal water mass sampled at Waterloo	Eurytemora affinis (*mean numbers/m³*)				Gammarus zaddachi (*mean numbers/m³*)			
	Surface	*Bottom*	*Day*	*Night*	*Surface*	*Bottom*	*Day*	*Night*
8	6.00	5.17	8.50	5.25	4.67	4.33	7.50	12.00
9	48.40	49.80	29.50	76.00	4.00	4.20	5.50	2.00
10	340.00	317.80	357.25	309.50	6.80	7.00	7.00	6.67
11	244.25	325.00	176.00	368.25	5.00	7.50	7.50	5.75
12	813.80	758.40	807.00	772.17	6.60	7.00	6.50	7.00

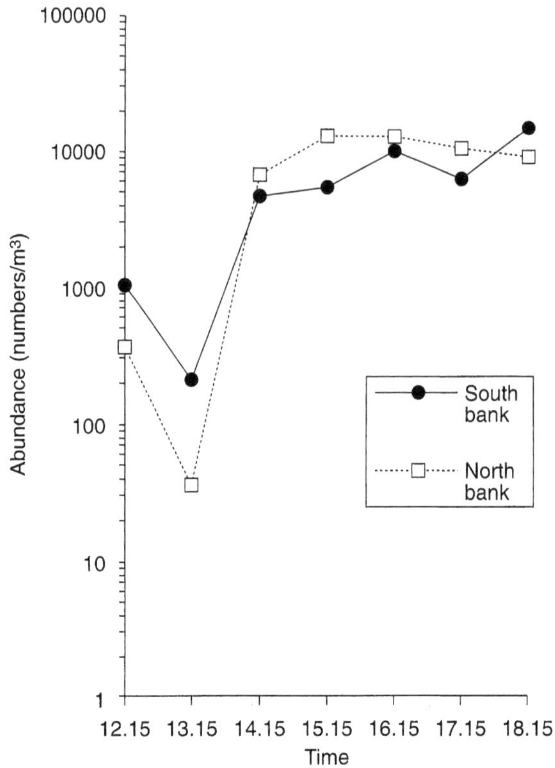

Figure 5.9 Abundance of *Eurytemora affinis* on the north and south banks of the Thames estuary at Blackwall, 30 April 1990.

been canalized for large stretches and has little marginal vegetation to act as refugia. Moreover, with the scaling-down of London as a port (Davies, 1969), many of the potential lateral habitats once afforded by docks and marinas are being lost.

5.7 COMPARISON WITH ZOOPLANKTON COMMUNITIES IN OTHER ESTUARIES

It is difficult to make detailed comparisons between estuarine zooplankton communities. This is because the dynamic nature of estuaries causes considerable variability, notably in patterns of salinity distribution. Moreover, different sampling methods can greatly influence the size and composition of zooplankton found. From the combined results of studies to date, it is apparent that the estuarine zooplanktonic fauna of the Thames is quite large, in regard to both species diversity and abundance of organisms, though many of the species present are meroplanktonic. In his study of the Forth estuary, Taylor (1987) identified a total of 135 taxa, which compares favourably with the Thames. Table 5.3 presents abundance estimates for the predominant crustacean *Eurytemora affinis* in the Thames and for other estuaries around the world. They are about the same order of magnitude (adult numbers present), apart from samples collected from the Bristol channel (Collins and Williams, 1981: Burkill and Kendal, 1982), where salinities were much closer to sea water. Soltanpour-Gargari and Wellershaus (1987) found *Eurytemora affinis* restricted to a very narrow range of low salinities in German estuaries, while Fulton (1984) found summer maxima of *Eurytemora affinis* in North America. In the Thames, *E. affinis* is found in waters with salinities approaching sea water and has a winter population maximum.

Quantitative assessment of the contribution of gammarids to zooplankton is difficult to find in the literature, presumably because its ability to occur in significant numbers in the water column has not previously been recognized. Mysids found in the macrotidal Conway estuary (Hough and Naylor, 1992) have a similar distribution pattern as is found in the Thames.

5.8 ZOOPLANKTON FUNCTIONING

5.8.1 Zooplankton feeding

The role of zooplankton in the biomanipulation of suspended solids has been well documented (Barthel, 1983; Herman and Scholten, 1990). Examination of the gut contents and faecal pellets of *Eurytemora affinis* by electron microscopy and by energy dispersive X-ray analysis indicate that the main dietary item is suspended sediment and detritus (Figure 5.10). Many studies indicate that detritus is the main, and in some cases the only, source of energy for estuarine zooplankton (Heinle and Flemer, 1975; Heinle *et al.*, 1977; Sellner and Bundy, 1987; Powell and Berry, 1990). The

Table 5.3 Abundance of *Eurytemora affinis* in selected estuaries

Estuary	*Abundance/m³*	*Source*
Bristol, England	100	Collins and Williams, 1981
Bristol, England	5	Burkill and Kendall, 1982
Weser, Germany	46 000	Soltanpour-Gargari and Wellershaus, 1984
Sacramento, USA	20 000	Orsi and Mecum, 1986
San Francisco Bay, USA	100 000	Ambler *et al.*, 1985
Thames, England	120 000	Lumkin, 1971
Thames, England	17 000	Present study

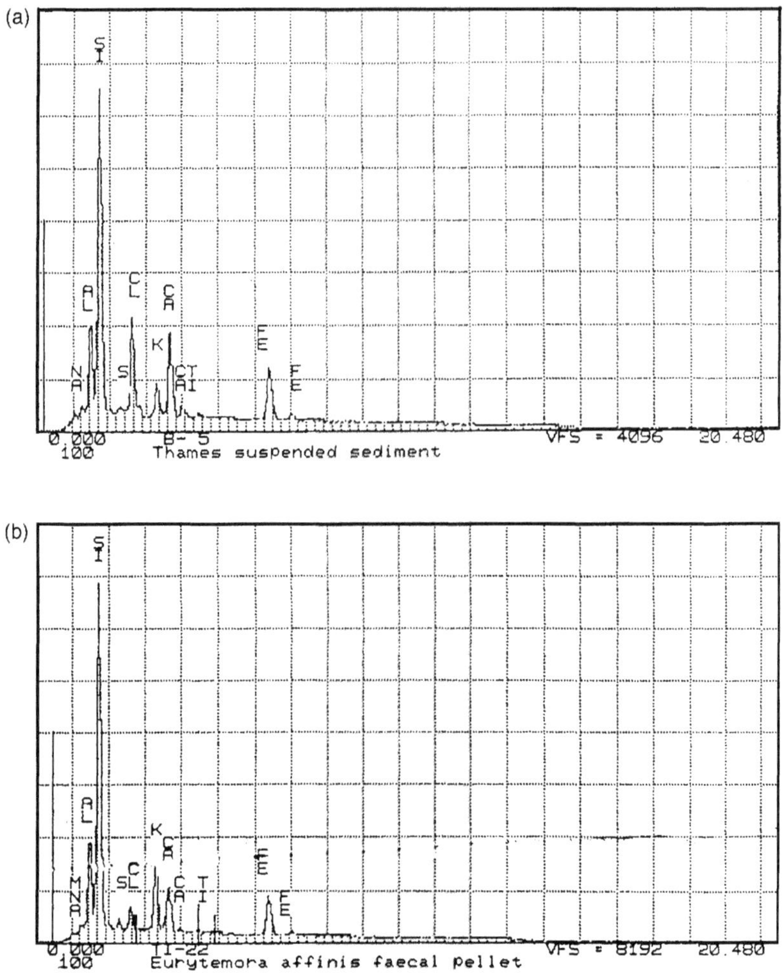

Figure 5.10 Elemental composition of **(a)** Thames suspended sediment and **(b)** faecal pellets of *Eurytemora affinis* by energy dispersive X-ray analysis.

situation in the Thames is made slightly more complex by the presence of large quantities of fibre, which reduces the feeding efficiency of some species such as *Eurytemora affinis* (Gordon *et al.*, 1993). The occurrence of these fibres had been observed by Trett *et al.* (1990) while sampling the meiofauna of the estuary. The fibre may originate from the sewage treatment process, where fibre is present as a breakdown product from toilet tissue. Dawson (1974) has noted large quantities of fibre around sewage treatment plant discharges in his studies on North American estuaries. An alternative source of the fibres may be discharges from the paper mills in the catchment of the estuary (Trevor Dean, personal communication).

5.8.2 Zooplankton as food organisms

Very little work has been performed on predator–prey relationships of organisms in the Thames. The most recent study (Mowah, 1991) presents some data on the gut contents of fish caught at the West Thurrock power station (Table 5.4). Mysids (mainly *Neomysis integer*) are preyed on by sand goby, bass, sprat, herring and sole. Gammarids (a mixture of *Gammarus salinus* and *G. zaddachi*) were favoured by five-bearded rockling and bass. The benthic *Corophium*, a fairly rare meroplanktonic organism, was eaten by sole, presumably from the estuary bed rather than in the water column. Surprisingly no copepods were recorded; in other estuaries, zooplanktonic copepods play a major role as food organisms for fish, and it is possible that these small organisms were overlooked in Mowah's study. Huddart and Arthur (1971b) reported that juvenile herring fed on *Eurytemora* in the Thames, as did sprat and three-spined stickleback. Sedgwick (1979) also reported sprat feeding on copepods in the Thames. *Eurytemora affinis* is the dominant prey of the delta smelt (*Hypomesus transpacificus*), and the second selected organism for larval striped bass (*Morone saxatilis*) in the upper Sacramento–San Joaquin estuary (Meng and Orsi, 1991; Moyle *et al.*, 1992).

Within the invertebrate community, several studies have implicated ctenophores in controlling the distribution of copepods in estuaries (e.g. Kremer and Kremer, 1988). In the Thames, the lowest numbers of some species coincided with high ctenophore populations, but there is insufficient quantitative evidence to show that this group markedly affects copepod numbers. Mysids also feed extensively on copepods (Fulton, 1982) and individuals of *Neomysis integer* were found in the Thames with fragments of *Eurytemora affinis* in their guts. Gaedke (1990) reported predation by *Noctiluca miliaris* on the eggs of *Eurytemora affinis* in the Elbe estuary. Egg predation was not noted in the Thames, and this might be the result of the spatial separation of these two species.

Table 5.4 Zooplankton from gut contents of fish caught at West Thurrock Power Station (after Mowah, 1991)

Scientific name	*Common name*	*Mysids*	*Corophium*	*Gammarids*
Ciliata mustela	Five-bearded rockling	+	+	+++
Pomatoschistus minutus	Sand goby	+++	–	+
Trispterus luscus	Bib	++	–	+
Merlangus merlangus	Whiting	+	–	+
Dicentrarchus labrax	Bass	+++	++	+++
Sprattus sprattus	Sprat	+++	++	–
Osmerus eperlanus	Smelt	++	–	+
Anguilla anguilla	Eel	+	–	++
Clupea harengus	Herring	+++	–	–
Solea solea	Sole	+++	+++	–

–, absent; +, rare; ++, occasional; +++, common

5.8.3 Other observations

In our study many adult individuals of *Eurytemora affinis* in the Thames were found covered by a heavy coat of epizoic ciliates *Epistylus* sp. Heckman (1986) reported the presence of the peritrich ciliate, *Myoschiston centropagidarum*, on *Eurytemora affinis* in the Elbe and Heerkloss *et al.* (1990) noted that epizoic ciliates increase mortality in crustacean nauplii stages.

5.9 CONCLUSIONS

From the results from our study, and from previous works on the Thames, it can be concluded that the zooplankton in the estuary has open-coastal, freshwater and resident estuarine components. Although holoplanktonic calanoid copepods dominate, a significant contribution is made by meroplanktonic amphipods. Amongst the copepods, the late-autumn to early-spring estuarine *Eurytemora affinis* and the summer euryhaline-opportunist *Acartia bifilosa* dominate the crustacean zooplankton of the estuary. Much remains to be discovered about the functioning of communities in specific regions of the tideway and especially the role of ctenophores, mysids and gammarids. The role of particulates as stressors for zooplankton needs to be assessed, as should the source of fibres in the water column. The zooplankton is clearly an integral part of the ecological system, with different communities associated with specific regions of the tideway, and much more attention should be devoted to its study.

CHAPTER 6

The benthic macroinvertebrate communities of the Thames estuary

Martin Attrill

6.1 INTRODUCTION

Over the past 20 years, the reports documenting the clean-up and recovery of the Thames estuary have concentrated largely on the return of the fish to the system, given their high public profile (Huddart and Arthur, 1971a,b; Doxat, 1977; Wheeler, 1979; Andrews, 1984). The return of the invertebrates has been placed somewhat in the background, especially those organisms living in and on the sediment ('benthic') covering both the intertidal and subtidal regions of the estuary bed. Considering most of the estuary's fish species rely on invertebrates as their primary food source (Wheeler, 1979), it could be stated that the fish would not have returned without the presence of established invertebrate communities. This chapter, together with Chapter 5, should help to restore the balance towards this important group of organisms.

Past work undertaken on the benthic macroinvertebrates (animals retained on a 0.5 or 1 mm sieve) in the Thames estuary has either been qualitative, concentrated on the distribution, biology and ecology of a single species, or described the organisms found at a particular location. The major previous report on the distribution of the macroinvertebrates was by Andrews *et al.* (1982), who described the macrofauna of the estuary (including fish), giving an invertebrate species list for the estuary as a whole obtained from observations and qualitative transects at a number of south-shore sites. Aston and Andrews (1978) concentrated on

A Rehabilitated Estuarine Ecosystem. Edited by Martin J. Attrill.
Published in 1998 by Kluwer Academic Publishers, London. ISBN 0 412 49680 1.

the fauna of London's freshwater rivers, including the upper reaches of the estuary, whilst Andrews (1977) documented changes in the distribution of organisms in this upper part of the Thames following the drought of 1976. More recently, studies resulting from the work of the National Rivers Authority (now Environment Agency) have furthered our understanding of the ecology of invertebrates within the Thames (e.g. Attrill *et al.*, 1996a,b; Attrill and Thomas, 1996).

Studies on individual species, or higher taxonomic levels, in the Thames have included oligochaetes (Birtwell, 1972; Hunter and Arthur, 1978: *Tubificoides* (*Peloscolex*) *benedeni*; Birtwell and Arthur, 1980: mainly *Tubifex costatus*), molluscs (Gibbs, 1993: *Nucella lapillus*) and crustaceans (Gee, 1961: *Corophium* spp.; Huddart and Arthur, 1971a: *Crangon crangon*), whilst some detail on the invertebrate community structure was provided at Erith (Andrews and Rickard, 1980) and at sites in the adjoining Medway estuary (Wharfe, 1977).

6.2 THAMES ESTUARY BENTHIC PROGRAMME

This chapter is based on the results from the first extensive, quantitative invertebrate survey to be undertaken along the whole length of the Thames estuary: the Thames Estuary Benthic Programme (TEBP). This was instigated in April 1989 by the National Rivers Authority (Thames Region) to survey the benthic invertebrate communities at a series of 28 sites from Teddington to the North Sea, covering both intertidal and subtidal areas. Appendix A includes full details of site locations, sampling methods, etc.

The survey concentrated on the macroinvertebrates, with quarterly sampling at each site, but meiofaunal surveys at each site were also undertaken (e.g. Trett *et al.*, 1990). The quantitative methods employed enabled both biomass (decalcified wet weight) and abundance to be attributed to each species for each sampling occasion, so by 1992 the data set obtained allowed a detailed picture of the structure of the communities present along the estuary to be constructed. The objective of this chapter is to present this picture.

6.3 THE THAMES DIVIDED

Chapter 2 described how management of the estuary is facilitated by dividing the whole length into a series of reaches, enabling objectives and plans to be constructed for each reach. The division of the estuary into zones can also be of benefit in terms of describing the macroinvertebrate communities present, if all sites within each zone exhibit a similar structure of species. This division can be achieved using the TEBP data set itself, by constructing categories relating to the similarities between the

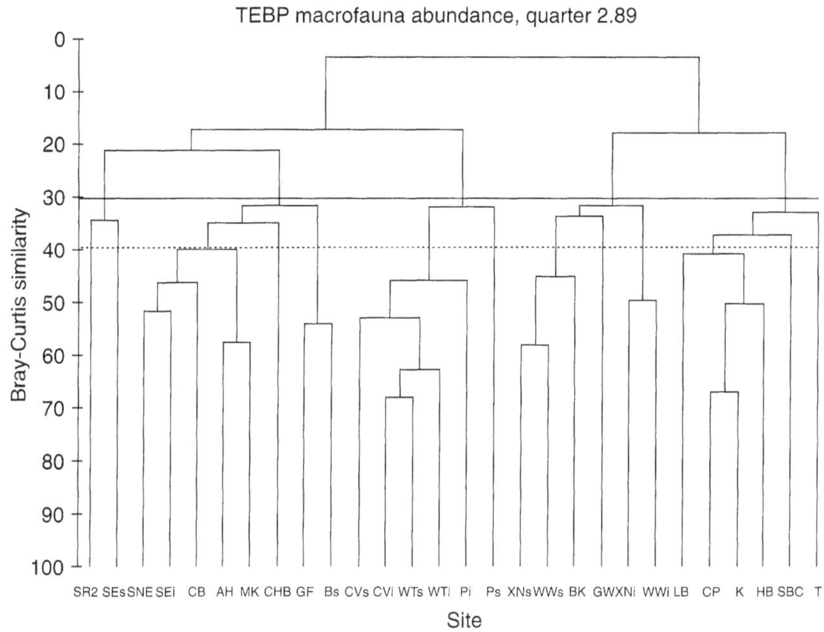

Figure 6.1 Cluster analysis of quarter 2.89 data from all sites using $\sqrt{\sqrt{}}$ transformed species abundances, Bray-Curtis similarity measures and group average sorting.

species complement at each site, the most similar sites being clustered together and consequently separated from sites with significantly different macroinvertebrate community structure.

Figure 6.1 illustrates a result of such multivariate analysis using a similarity matrix, the dendrogram relating the sites to each other in terms of their degree of similarity. This relationship can be graphically presented using ordination techniques (Figure 6.2), with each site being plotted out in two-dimensional space, the distances between sites relating to their varying degrees of similarity. By choosing a level of similarity from Figure 6.1 which gives meaningful results, e.g. 30%, clusters can be constructed on the ordination relating to sites with greater or less than 30% similarity (for more detail on the techniques used see Clarke and Warwick, 1994).

In terms of its macroinvertebrate community structure, the Thames has now been divided into five zones at a 30% similarity level, with a general trend of increasing salinity from zones 1 to 5 (cf. Chapter 4). The data set chosen for this classification was obtained in the second quarter (April–June) of 1989, which was the first sample period during the TEBP. Until the end of this period, the Thames had been experiencing a typical flow regime, the last major drought occurring in 1976. The clustering of sites can therefore be characterized as the 'expected' order under such

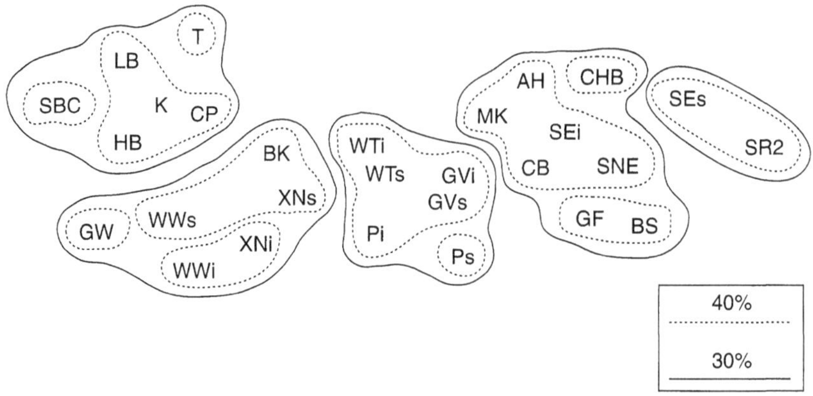

Figure 6.2 MDS ordination of quarter 2.89 abundance data using the same transformations and similarity measures as Figure 6.1, with the groupings from the dendrogram superimposed. Stress = 0.16.

conditions and so provides an ideal basis for the description of macroinvertebrate communities in terms of zones. Any variation on this community structure can then be related to changes in physicochemical conditions, particularly those associated with changes in river flow (Chapter 3; Attrill *et al.*, 1996a).

6.4 BENTHIC MACROINVERTEBRATES – ZONE BY ZONE

6.4.1 Zone 1: Teddington to London Bridge

This upper zone is dominated by the influence of freshwater entering the estuary over Teddington Weir; consequently, the communities at the sites within the zone were characteristically composed of freshwater invertebrate species. However, the diversity of this group of species diminished rapidly away from the weir, leaving a particular group of salinity-tolerant species to provide the basis of the community structure for the estuary downstream of Teddington.

The vast majority of freshwater organisms are intolerant of exposure to air, and so diagnostic species comprising the core community are to be found on the stone/shingle substrate of the estuary channel below the low tide mark (Figure 6.3). The dominant species in terms of both abundance and biomass was the amphipod *Gammarus zaddachi* (Figure 6.4), an upper estuarine rather than strictly freshwater species, and the pivotal species in terms of the ecology of the upper Thames estuary. Its large numbers and palatability make it the most important single food species for the majority of fish species (Wheeler, 1969) and the vector for the transmission

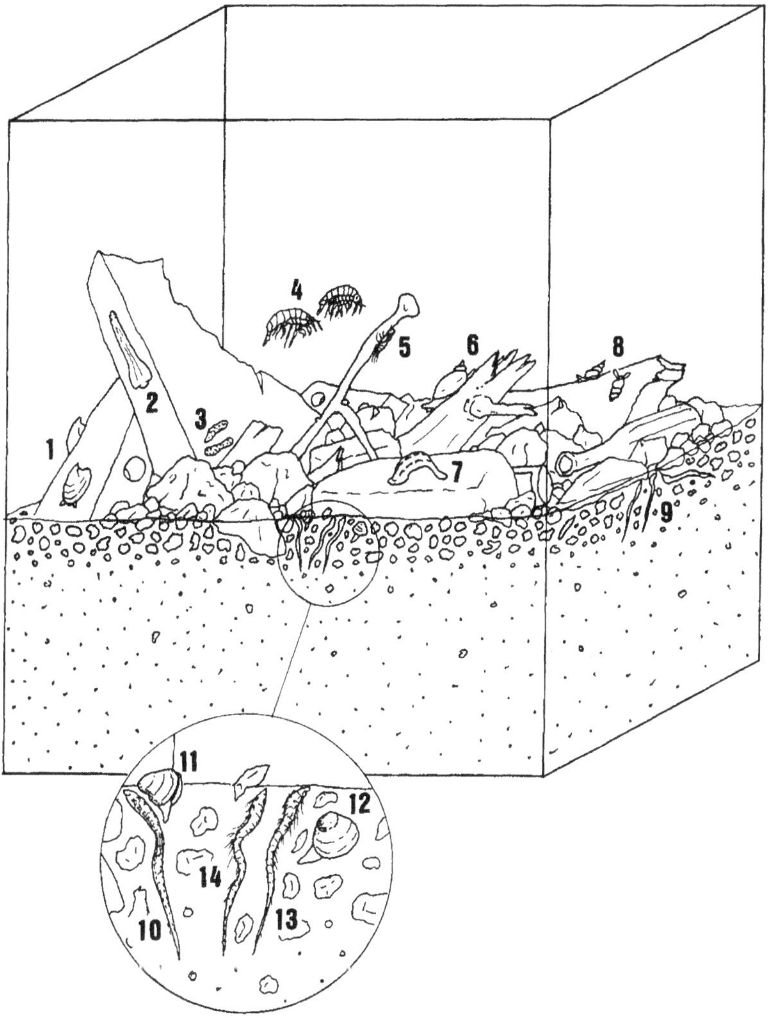

Figure 6.3 Stylized diagram of the core community present at Kew. **1** *Ancylus fluviatilis*; **2** *Erpobdella testacea*; **3** *Lymnaea peregra* egg masses; **4** *Gammarus zaddachi*; **5** *Caenis moesta*; **6** *Lymnaea peregra*; **7** *Glossiphonia complanata*; **8** *Potamopyrgus jenkinsi*; **9** Oligochaeta spp.; **10** *Limnodrilus hoffmeisteri*; **11** *Pisidium* sp.; **12** *Sphaerium corneum*; **13** *Psammoryctides barbatus*; **14** *Tubifex tubifex*.

of certain fish parasites (Chapter 8), as well as having a major impact on the status of prey species. The abundance of *G. zaddachi* varied markedly, with the maximum number (taken in a three-minute kick sample) being 7250 at Kew on 24 June 1990. On the previous sampling occasion (15 May 1992) only 138 were present. The amphipod was only ever absent in one set of samples, at Teddington, on the same day that 7250 individuals were

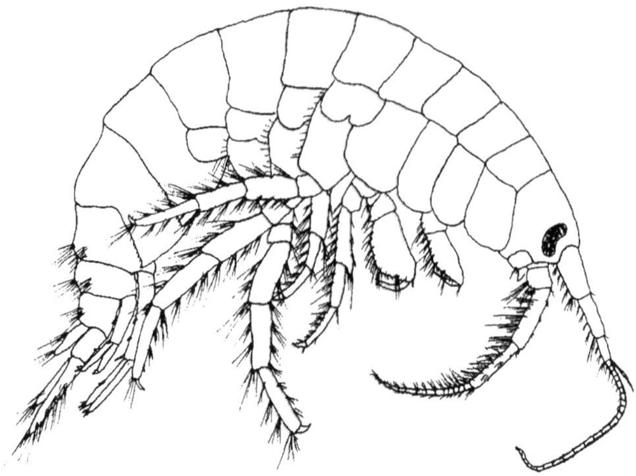

Figure 6.4 *Gammarus zaddachi* (after Lincoln, 1979).

present at Kew. This followed a period of very high flows, appearing to wash the *G. zaddachi* downstream. On this occasion, the *G. zaddachi* at Teddington were replaced by a few individuals of *Crangonyx pseudogracilis* and *Gammarus lacustris*, a species normally found in still waters.

Of the other core species in the Zone 1 communities, molluscs and oligochaetes are the best represented. Three mollusc species were particularly prevalent: the river limpet *Ancylus fluviatilis*, the wandering snail *Lymnaea peregra* and the small hydrobiid gastropod *Potamopyrgus jenkinsi*. All appear tolerant of a small increase in saline conditions, although *P. jenkinsi* was originally a brackish water species that only invaded freshwater systems in the late nineteenth century (Macan, 1969) and consequently is found in the estuary below Zone 1. Snails at each end of the salinity range show marked morphological differences, however. Individuals inhabiting the freshwater reaches of the Thames above and directly below Teddington weir tend to be small, smooth and dark (Figure 6.5a), while individuals found further down the estuary (e.g. Greenwich) are much larger, lighter and often have a pattern of ridges, keels and frills (Figure 6.5b). The transition point appears to be around the Hammersmith area, where the snail was regularly absent. It is possible that populations of the species have shown some divergence since penetrating freshwater, and further genetic investigation of *P. jenkinsi* would be interesting.

Bivalve species are represented by the Sphaeriidae (*Sphaerium corneum*, *Pisidium* spp.) in the upper half of Zone 1, together with the large swan mussel *Anodonta complanata*. This species is present in considerable numbers in the deeper parts of the channel in the upper estuary, but due

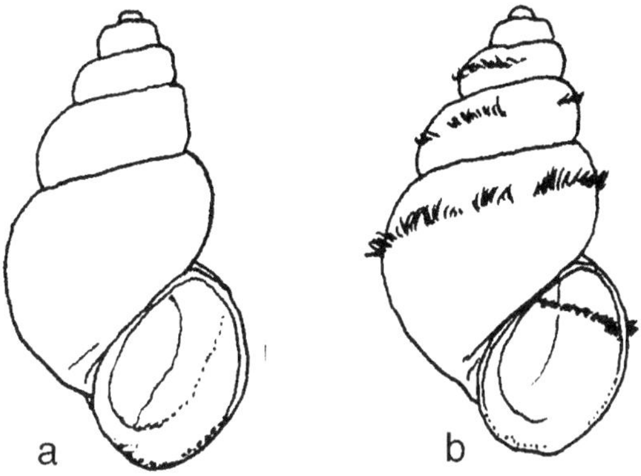

Figure 6.5 *Potamopyrgus jenkinsi*: **(a)** smooth type; **(b)** type with keel.

to its habit is rarely encountered. The lock at Richmond maintains the level of water in the estuary above this point at an artificially high level during the low-tide period (Chapters 2 and 3; Bowen and Pinless, 1977). On rare occasions when the lock is not operating (maintenance, etc.) an unusual natural low tide occurs above Richmond, exposing more extensive areas of estuary bed together with large numbers of *A. complanata* stranded by the atypical water levels.

Over 20 species of oligochaete have been recorded from Zone 1, the majority being members of the Tubificidae, though only a few are widespread and common enough to be considered core species. These are (in a general order of degree of ubiquity) *Psammoryctides barbatus*, *Limnodrilus hoffmeisteri* and *Tubifex tubifex*. In addition, *Limnodrilus cervix*, *Potamothrix hammoniensis* and members of the Naididae (particularly *Nais elinguis*) were regularly encountered. *Limnodrilus claparedianus* appears to be particularly common around the Battersea area, with *Aulodrilus pluriseta* forming a localized population at Teddington. It would appear that this latter species is particularly intolerant of salinity increases.

A further annelid group with widespread representatives in Zone 1 is the leeches (Hirudinea). Members of both the Glossiphoniidae and Erpobdellidae were commonly encountered, with *Erpobdella testacea* having been recorded at all sites. Glossiphoniidae are more restricted in their downstream distribution, but both *Glossiphonia complanata* and *Helobdella stagnalis* were frequently recorded in the estuary above Kew. *Dendrocoelom lacteum*, a member of a more primitive group, the flatworms (Platyhelminthes), was regularly encountered in the upper half of the zone, particularly under larger rocks. Figure 6.3 presents a stylized picture of the core community found within Zone 1, using the faunal structure at Kew as an example. The core species featured in Figure 6.3 exhibited different distributions along the Thames estuary, depending on their degree of tolerance to increasing saline conditions. For example, neither *Lymnaea peregra* nor *Ancylus fluviatilis* was recorded at London Bridge, where the increased level of salinity is exacerbated by the harsh physical conditions of a coarse, mobile sediment and high tidal flows.

Of all the groups of freshwater taxa recorded in the estuary, the insects appear the least tolerant to any increase in salinity and were generally confined to the Teddington area. The only widespread species were the mayfly *Caenis moesta*, which was recorded across the whole zone following the high flows at the start of 1990, and larvae of the Chironomidae, a family that includes several salinity tolerant species, e.g. *Chironomus* and *Thalassosmitia* spp. (Eppy, 1989). However, the estuary just below the weir at Teddington recorded a wide diversity of insect species. These included caddisfly larvae (*Anthripsodes cinereus, Ceraclea nigronervosa, Mystacides longicornis, Tinodes waeneri*), mayfly larvae (*Cleon dipterum, Caenis horaria, Ephemera danica, Baetis* sp.) and beetle larvae and adults (*Oulimnius tuberculatus, Dytiscus* sp., *Deronectes depressus*).

The intolerance of many of these freshwater species to even small increases in saline conditions was demonstrated during the periods of drought in the summers of 1989 and 1990. To replenish dwindling drinking water supplies, water abstraction from the river Thames between Windsor and Teddington was increased, resulting in a decrease in freshwater flows entering the estuary over Teddington Weir. This decrease in flows was accompanied by an increase in salinity levels at sites in the Thames estuary (Attrill *et al.*, 1996a). The changing conditions had a profound effect on the structure of the benthic macroinvertebrate community within Zone 1, particularly at the site situated below Teddington Weir. Figure 6.6 illustrates the relationship between small increases in salinity and species number recorded at the site. During the low-flow periods, most of the more salinity-sensitive species detailed above disappeared, the species remaining being those core species tolerant of salinity that are prevalent at the other sites within Zone 1 (Figure 6.3). As flows increased again, the site at Teddington was 'reseeded' with freshwater invertebrates drifting over the weir from the river.

The decrease in flows also had a notable effect on the lower half of Zone 1. Several estuarine species showed an upstream movement following the saline incursion and were regularly recorded in samples from sites below Kew. These included the estuarine prawn *Palaemon longirostris*, brown shrimp *Crangon crangon* and the mysid *Neomysis integer*. Smaller, less mobile estuarine organisms also showed an upstream extension of their range during the drought periods. Most notable was the amphipod *Corophium lacustre*, forming large numbers of burrows between London Bridge and Battersea and recorded as far upstream as Kew. The polychaete *Polydora* sp. was recorded as far up as Hammersmith, while the isopod *Sphaeroma rugicauda* was present at Cadogan Pier, Battersea.

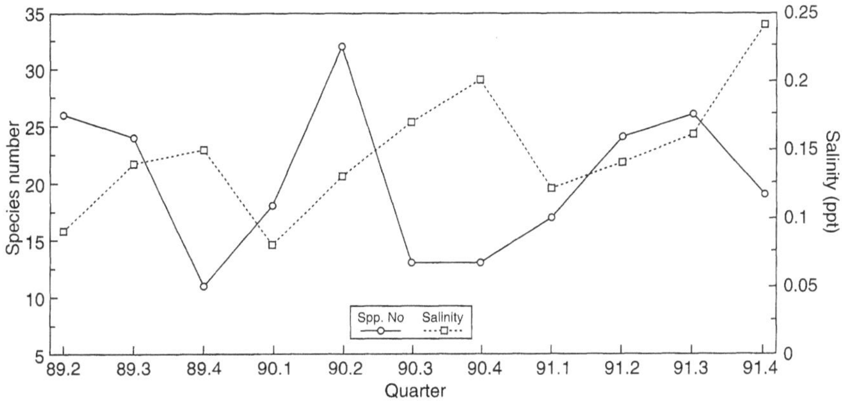

Figure 6.6 Species number and mean quarterly salinity (mid-tide corrected) recorded at Teddington, 1989–1991.

Areas of soft sediment in Zone 1 are represented by a few intertidal sand banks that have developed on the inside of some bends, e.g. at Hammersmith Bridge and the South Bank Centre. The fauna present in these sandy areas is very limited, as might be expected, and dominated by the oligochaete species detailed earlier, generally in low numbers. This base of oligochaete species is supplemented by small *Gammarus zaddachi*, together with occasional small individuals from other core species (e.g. *Sphaerium corneum*, *Potamopyrgus jenkinsi* and Chironomidae spp.). The maximum number of species recorded in the sand at Hammersmith was 14, including eight oligochaete species.

Due to the zone's inherent freshwater nature, many species recorded from Zone 1 were not recorded elsewhere in the estuary, particularly the freshwater fauna under the weir at Teddington (as mentioned earlier). However, the zone also contained some interesting species confined to other sites, the most unusual being a population of the large tubificid oligochaete *Branchiura sowerbyi* present at Kew. This is essentially a warm-water species (Brinkhurst, 1971), the population at Kew being particularly prevalent in the summer of 1989. A further species recorded only at Kew was the gastropod *Viviparus viviparus*. This snail is present in the freshwater Thames above Teddington weir, but not in the estuary a few hundred metres further down, appearing to prefer the underside of large boulders. In recent years, young Chinese mitten crabs (*Eriocheir sinensis*) have become a feature of the intertidal reaches along this part of the Thames, these small crabs migrate upriver into freshwater, where they live as adults. This introduced species has only recently established a breeding population, despite being sporadically recorded in the estuary for over 50 years (Attrill and Thomas, 1996).

6.4.2 Zone 2: London Bridge to Crossness

Zone 2 is characterized by the appearance of the first permanent, though often highly mobile, intertidal and shallow subtidal mudflats. Small patches of mud occur at Greenwich, and by Woolwich the intertidal areas are quite extensive. Under normal flow conditions, the mid-tide salinity of this region is still generally low (5–10 PSU) and represents the transition area from freshwater to estuarine conditions. The two large sewage treatment works also discharge into this zone at Beckton and Crossness, the combination of physicochemical factors making Zone 2 probably the most stressful environment for benthic macroinvertebrates in the estuary. The result is a limited fauna comprising freshwater species that can tolerate the increased salinity and estuarine species capable of withstanding wide variations in the saline regime.

The intertidal mudflats were dominated by tubificid oligochaetes, principally the freshwater species *Limnodrilus hoffmeisteri* and the estuarine species *Tubifex costatus*. *L. hoffmeisteri* was the dominant species present in the patches of intertidal mud at Greenwich, where it has been recorded in densities of over 5000/m^2, and at Woolwich, though in lower numbers. *T. costatus* has been regularly recorded at both sites, but appears to be restricted by the low salinity. However, the extensive, comparatively stable mudflats near the sewage outfall at Crossness seem to provide the optimum environment for this species. Here *T. costatus* has been recorded in densities higher than any other intertidal species in the Thames estuary, reaching a peak of 27 065/m^2 in the fourth quarter of 1990. This resulted in one of the highest biomass totals for the estuary (Table 6.1). *L. hoffmeisteri* was also numerous at Crossness, peaking at over 10 000/m^2 in the second quarter of 1990, although mature specimens were only rarely encountered. Consequently, the intertidal mudflats off Crossness appear to be the only remaining area in the Thames estuary solely dominated by vast numbers of tubificid oligochaetes, a remnant of past conditions across a large area of the mid-estuary (Chapter 1).

The two main oligochaete species are supplemented by a comparatively restricted suite of organisms that can be regarded as characteristic for the intertidal areas within Zone 2. These core species include the amphipods *Gammarus zaddachi* and *Corophium lacustre*, the gastropod *Potamopyrgus jenkinsi* and the oligochaetes *Tubifex tubifex* and *Monopylephorus rubroniveus*. Two classic estuarine species, *Corophium volutator* and *Nereis diversicolor*, have been recorded regularly at Crossness in low abundances, with occasional records upstream. In addition, the shore crab *Carcinus maenas* has been observed on the shore as far up as Woolwich. The influence of the freshwater input from the sewage outfall on the fauna of the Crossness mudflat is indicated by the records of two freshwater oligochaete species (*Lumbriculus variegatus*, *Limnodrilus udekemianus*) that have not been recorded at Zone 2 sites above Crossness.

The influence of lower freshwater flows was apparent at the intertidal sites within Zone 2, with several species showing an upstream increase in

Table 6.1 Sites in the Thames estuary recording the highest total biomass

Site		*Quarter*	*Biomass (gWW/m²)*	*Dominant species*
1	Subtidal at Canvey (3m)	2/89	1085.12	*Crepidula fornicata*
2	West Thurrock (intertidal)	3/89	769.20	*Nereis diversicolor*
3	Chapman Buoy (subtidal 20m)	4/89	253.30	*Sagartia troglodytes*
4	West Thurrock (subtidal 2m)	2/89	150.62	*Corophium volutator*
5	Gravesend (intertidal)	2/91	149.64	*Corophium volutator*
6	Allhallows (intertidal)	2/89	130.24	*Mytilus edulis*
7	Southend (intertidal)	4/90	94.30	*Cerastoderma edule*
8	Canvey Beach (intertidal)	1/91	72.83	*Mytilus edulis*
9	Crossness (intertidal)	4/90	71.29	*Tubifex costatus*
10	Shoeburyness E (intertidal)	3/89	67.60	*Scoloplos armiger*

their range. The upper records have included the bivalve *Macoma balthica* (Crossness, 4/90), the catworm *Nephtys hombergi* (Woolwich, 3/90) and oligochaete *Tubificoides benedeni* (Woolwich, 2/91), an upstream range extension comparable with that for certain algae species (Chapter 4).

The subtidal areas within this zone exhibit a slightly different invertebrate structure, with a more variable community in terms of both species composition and abundance, influenced by the highly mobile nature of the sediment. The mudflats extend into the shallow subtidal, before the scouring of the mid-channel region results in little but bare stones and gravel. A notable exception to this structure is present in the shallow subtidal region off Greenwich. Here the community is based on pebbles partially embedded in the mud, resulting in a stable hard substrate (Figure 6.7). These pebbles are covered in the mud burrows of the amphipod *Corophium lacustre,* the burrows in turn providing a habitat for other invertebrates. It is practically impossible to assess the density of the *Corophium* at this site quantitatively: the pebbles jam in the jaws of the grabs, resulting in partial samples. However, counts made from these partially washed samples alone have yielded *Corophium lacustre* densities of > 80 000/m². The *C. lacustre* population has an associated community (Figure 6.6). The secured pebbles provide a substrate for large numbers of the zebra mussel, *Dreissena polymorpha.* The growth of these introduced bivalves appears stunted, with few developing the characteristic markings. *Potamopyrgus jenkinsi* and *Gammarus zaddachi* were both common amongst the *Corophium* burrows, together with the isopod *Sphaeroma rugicauda.* Beneath the pebbles, the mud supports oligochaetes – particularly *Limnodrilus hoffmeisteri* and *Psammoryctides barbatus* – while empty burrows provide habitats for the spionid polychaetes *Streblospio shrubsolii* and *Polydora* sp.

The mobile larger crustaceans present at Greenwich (*Palaemon longirostris, Crangon crangon, Neomysis integer*) were common throughout

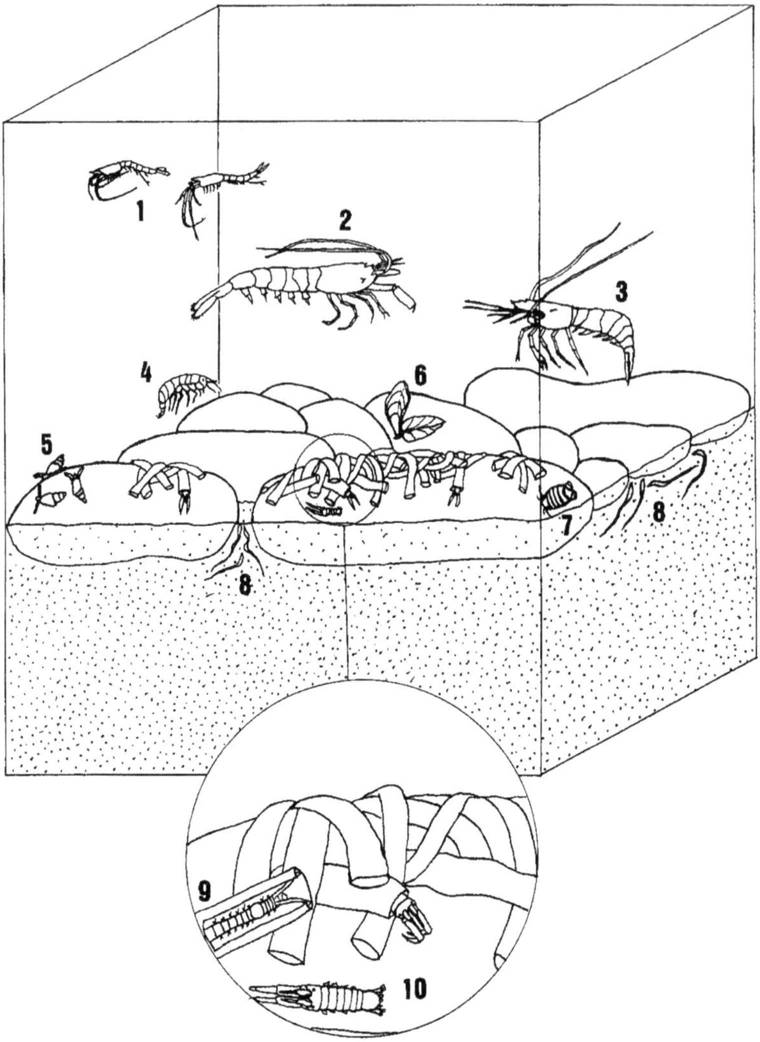

Figure 6.7 Stylized diagram of the 'pebble' community present at Greenwich. **1** *Neomysis integer*; **2** *Crangon crangon*; **3** *Palaemon longirostris*; **4** *Gammarus zaddachi*; **5** *Potamopyrgus jenkinsi*; **6** *Dreissena polymorpha*; **7** *Sphaeroma rugicauda*; **8** Oligochaeta spp. (*Psammoryctides barbatus*, *Limnodrilus hoffmeisteri*); **9** *Polydora* sp.; **10** *Corophium lacustre*.

the subtidal region of Zone 2. Seawards of Greenwich, the benthic invertebrates become less abundant, though *Limnodrilus hoffmeisteri* remains one of the most common species. However, it was replaced as the dominant oligochaete species by the small tubificid *Monopylephorus rubroniveus*, which appears to thrive in the difficult, mobile conditions experienced in the subtidal reaches of Zone 2, reaching densities of up to 500/m^2 at Woolwich and Crossness. Polychaete species were also more prevalent in the subtidal areas than the intertidal flats, with *Streblospio shrubsolii*, *Nereis diversicolor* and *Polydora* sp. regularly recorded. Below Greenwich, *Corophium lacustre* is replaced by *Corophium volutator*, though in Zone 2 this species rarely forms large populations, the maximum density recorded being 300/m^2 at Crossness (2/90). The *Dreissena* growing on subtidal boulders continue (though in progressively lower numbers) from Greenwich to Woolwich, where the barnacle *Balanus improvisus* begins to colonize subtidal hard surfaces.

The subtidal site adjacent to Beckton sewage works was notable as being the poorest site in terms of benthic macroinvertebrates in the Thames estuary. Despite having very similar sediment characteristics to Woolwich and Crossness, the site consistently recorded the lowest number of species (Figure 6.8). During five quarters, the only species present was *Limnodrilus hoffmeisteri*, usually in low numbers, while during the third quarter of 1990, no macroinvertebrates were present in any of the subsamples. This was the only occasion when a 'lifeless' site was recorded during the TEBP. The majority of other organisms recorded here were mobile crustaceans, such as *Gammarus zaddachi*, *Corophium volutator* and *Neomysis integer*. During quarter 2/91, *G. zaddachi* was replaced by *Gammarus salinus*, following an upstream movement of this species. A further upstream record in Zone 2 was for the polychaete *Nereis succinea*, present in the Crossness subtidal samples during the low flow period at the end of 1989.

Few of the species recorded from Zone 2 were particularly unusual or unique to the zone, the only singular occurrence being a larva of the phantom midge *Chaoborus* sp. recorded from the intertidal mudflats at Woolwich during quarter 1/91. On 11 June 1991, three individuals of the freshwater gastropod *Physa heterotropha* were recorded at Crossness (together with two further downstream at West Thurrock). All were alive, their appearance in such comparatively high salinity conditions being somewhat of an anomaly.

6.4.3 Zone 3: Crossness to Gravesend

This second mid-estuary zone covers the transition from estuarine to more marine conditions, with mid-tide salinity within the range 10–20 PSU. The sediment characteristics within Zone 3 are similar to the previous zone, with extensive areas of intertidal mudflat and rather unstable regions of subtidal soft sediment. However, with increasing salinity within Zone 3 there is an increase in the sand content of the substrate, giving a range from fine mud at Purfleet to sandy mud at Gravesend. The increasing salinity is reflected in the fauna of the zone, which is characterized by a mixture of estuarine and marine species, many of them having their upstream limit in Zone 3.

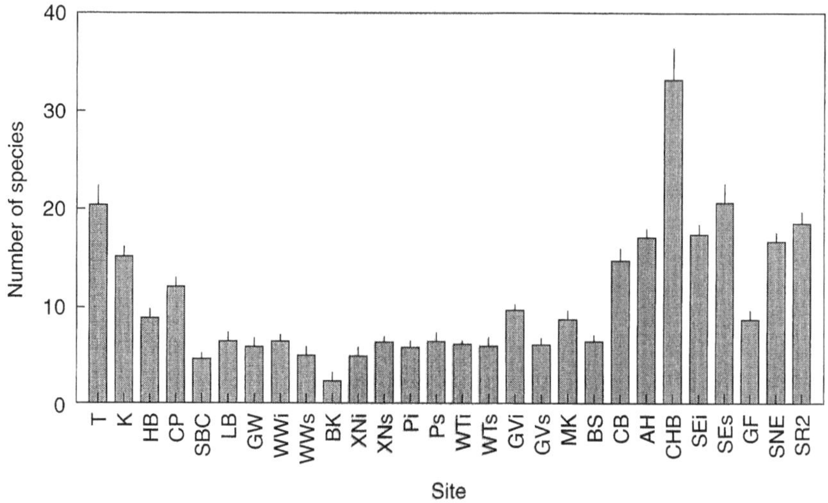

Figure 6.8 Mean species number recorded at each site along the length of the Thames estuary, together with standard deviations.

Like Zone 2, the intertidal flats of this zone recorded high numbers of tubificid oligochaetes. However, they are accompanied in Zone 3 by other, larger abundant species. *Tubifex costatus* remains the most common oligochaete in the upper half of the zone, with densities over 7500/m^2 at West Thurrock, but this is gradually replaced by *Tubificoides benedeni* (Figure 6.9), the dominant species in terms of abundance at Gravesend, where > 7000/m^2 have been recorded. Unlike the situation at Crossness, oligochaetes rarely contribute a significant amount towards the biomass at the two sites. The intertidal flats at West Thurrock constitute one of the most productive habitats in the Thames (Table 6.1) due to high numbers of *Corophium volutator* and particularly of the ragworm *Nereis diversicolor*, the polychaete having provided a biomass of over 760 g wet weight (WW)/m^2 (quarter 3/89) and densities of over 4600/m^2 (quarter 2/90). *Tubifex costatus*, *C. volutator*, *N. diversicolor* and low numbers of *Tubificoides benedeni* constituted the core community at West Thurrock, supplemented by a few other less frequent species such as *Gammarus zaddachi*, *Carcinus maenas* and the bivalves *Macoma balthica* and *Scrobicularia plana*.

The dominant species at Gravesend, in terms of biomass, was generally *Corophium volutator*, although its numbers were highly variable. On 3 June 1991 a total of 8825 *C. volutator*/m^2 was recorded – the highest figure for an intertidal site in the Thames. However, the next sample (2 September 1991) recorded a density of only 5/m^2. Due to the increase in salinity regime, the Gravesend intertidal area generally recorded a wider range of species present than West Thurrock. Core species regularly occurring here include the polychaetes *Nephtys hombergi* and *Caulleriella* sp., the bivalves

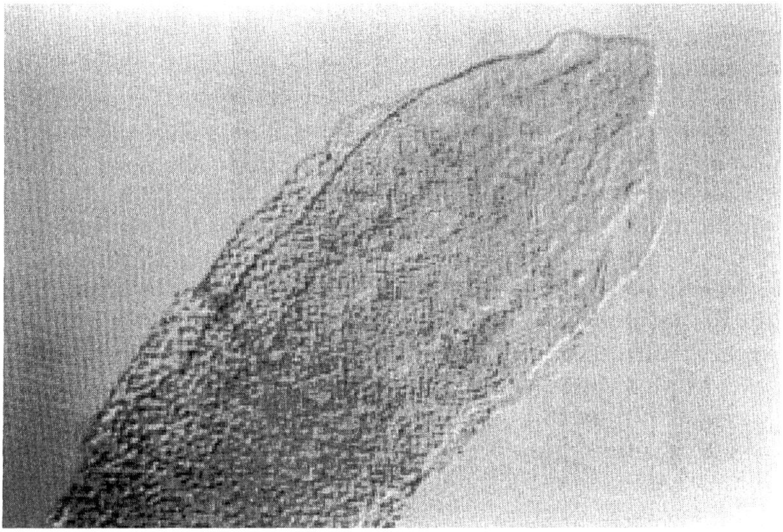

Figure 6.9 *Tubificoides benedeni*, anterior end.

M. balthica and *S. plana* and the amphipod *Gammarus salinus*. The transition between *G. zaddachi* and *G. salinus* appears to be between West Thurrock and Gravesend under normal flow conditions. Gravesend is the upstream limit for the classic intertidal species *Littorina littorea* and *Arenicola marina*.

At the upper end of Zone 3, the intertidal mudflats at Purfleet supported a poor, low abundance fauna compared with West Thurrock and Gravesend. Most of the core species present at the lower two sites have been recorded at Purfleet, but more erratically, due to the instability of the sediment and the lower, more variable salinity regime in this part of the estuary. The site generally recorded a total biomass of < 1 g WW/m^2. Oligochaetes (e.g. *Limnodrilus hoffmeisteri*, *Tubifex costatus*, *Tubificoides benedeni*) were usually present in low numbers, together with small polychaetes (spionids, young *Nephtys* and *Nereis*) and occasional Crustacea (generally *Corophium volutator*). In the second quarter of 1990, large numbers of mobile crustaceans (*Crangon crangon*, *Neomysis integer*, *Palaemon longirostris*) were present in the mud, resulting in the highest biomass for the site (8 g WW/m^2). The intertidal area at Purfleet also provides one of the only rocky shores (i.e. large, immobile boulders and concrete) in the Thames estuary, but due to the nature of the salinity regime few organisms have colonized it. The main species present were estuarine crustaceans such as *Carcinus maenas*, *Jaera albifrons* gp., *Balanus improvisus* and *Sphaeroma rugicauda*.

The subtidal regions of Zone 3 demonstrated a similar pattern to Zone 2, with a highly variable, patchy fauna influenced by sediment mobility. However, stable patches of substratum within this zone appear to be able to support diverse, productive communities. Subtidal sites at Purfleet, West Thurrock and Gravesend all demonstrated comparatively rich communities, but vast changes in abundance and structure between samples highlighted the variability. The trend in species number and abundance appeared to follow changes in the abundance of *Corophium volutator*, the highest species numbers at all sites coinciding with high numbers of the amphipod. Other core species indicative of the Zone 3 subtidal areas are the polychaetes *Streblospio shrubsolii*, *Nephtys hombergi*, the brown shrimp *Crangon crangon* and three oligochaete species: *Monopylephorus rubroniveus*, *Tubifex costatus* and *Tubificoides benedeni*. Only *T. benedeni* was ever continually present (at Purfleet), due to the high degree of variability from one sample to the next. The most dramatic example of this was at West Thurrock in 1989 (Table 6.2), the population of *C. volutator* being reduced from over 20 000 to 10/m^2 between two samples, with a corresponding disappearance of all other species. The macroinvertebrate crash coincided with a similar depletion of the meiofaunal community, with only the nematode *Sabatiera punctata* remaining.

Similar changes in community structure, though less dramatic, were noted for both Purfleet and Gravesend. The crash at Gravesend also occurred between quarters 2/89 and 3/89, but from the end of 1989 the fauna here tended to be very poor, with generally low species numbers and a total biomass of < 1 g WW/m^2. Only a single *Nephtys hombergi* was recorded in four Day grabs taken on 26 September 1990. There is a very rapid transition in this part of the estuary from intertidal mud to the deep channel, giving a very narrow band of subtidal soft sediment. This appears to be constantly disturbed, preventing any degree of settlement. The change from the high abundance intertidal community at Gravesend to this species-poor, variable subtidal habitat occurs across the low tide mark – a horizontal distance of < 20 m.

The subtidal regions of Zone 3 registered the upper limit for several marine/outer estuarine organisms, many of these species recording an upstream movement during the periods of low freshwater flow. These included the polychaetes *Pygospio elegans*, *Caulleriella* sp., *Eteone longa* and *Spio filicornis* (all recorded up to Purfleet), the amphipod *Melita obtusata* (West Thurrock), the molluscs *Hydrobia ulvae* (Purfleet) and *Mytilus edulis* (West Thurrock), the sea squirt *Molgula manhattensis* (West Thurrock) and the cnidarians *Tubularia indivisa* and *Sagartia troglodytes* (Gravesend).

The only species uniquely recorded in Zone 3 was the prawn *Palaemonetes varians* at Purfleet intertidal on 26 September 1990. However, many large benthic and nektonic invertebrate species (e.g. cephalopods, jellyfish, crabs, prawns, etc. which do not occur in grab samples) are entrapped on the cooling water intake screens at West Thurrock Power station during fish surveys. Chapter 7 describes this sampling method in full detail, with invertebrate species recorded here included in the species list in Appendix B.

Table 6.2 Changes in the subtidal macroinvertebrate community at West Thurrock during 1989 (expressed as abundance/biomass)

Species	*Quarter 2/89*	*Quarter 3/89*	*Quarter 4/89*
Oligochaeta			
Tubifex costatus	772.5/1.50		162.5/0.20
Tubificoides benedeni	800/1.53		85/0.15
Polychaeta			
Caulleriella sp.			12.5/0.03
Nephtys hombergi			2.5/0.05
Nereis diversicolor	40/1.70		
Nereis succinea	10/0.98		107.5/8.93
Streblospio shrubsolii			100/0.18
Crustacea			
Corophium volutator	22125/144.78	10/0.10	362.5/3.35
Crangon crangon	2.5/0.08		
Gammarus zaddachi	2.5/0.05		
Totals	23752.5/150.62	10/0.10	845/14.09
Total No. Species	7	1	8

Abundance = no. individuals/m^2
Biomass = g WW/m^2 (WW = wet weight)

6.4.4 Zone 4: Outer Estuary I

Zone 4 encompasses most of the outer Thames estuary, including all the extensive intertidal mud and sand flats and the subtidal areas between Gravesend and Canvey Island. The variety of macroinvertebrate communities present in this zone can be best illustrated by subdividing the zone into two sections: firstly, the northern intertidal flats from Canvey to Shoeburyness East; and secondly, a transect across the Thames from Canvey to Allhallows.

(a) The north shore

Vast areas of mud and sand are exposed at low tide along the length of the north bank of the outer Thames estuary. These range from the 100–200 m of sandy mud revealed off Canvey Island to the expansive sand flats of Foulness and Maplin Sands, where the tide retreats up to 6 km out into the North Sea. These intertidal areas are internationally important for wintering wildfowl and waders, being supported by the extremely productive macroinvertebrate and meiofauna communities present. There is a general trend of increasing biomass from Canvey to Shoeburyness East, together with a greater depth-penetration of the substrate by organisms as the sediment becomes more sandy. Figure 6.10 illustrates the core

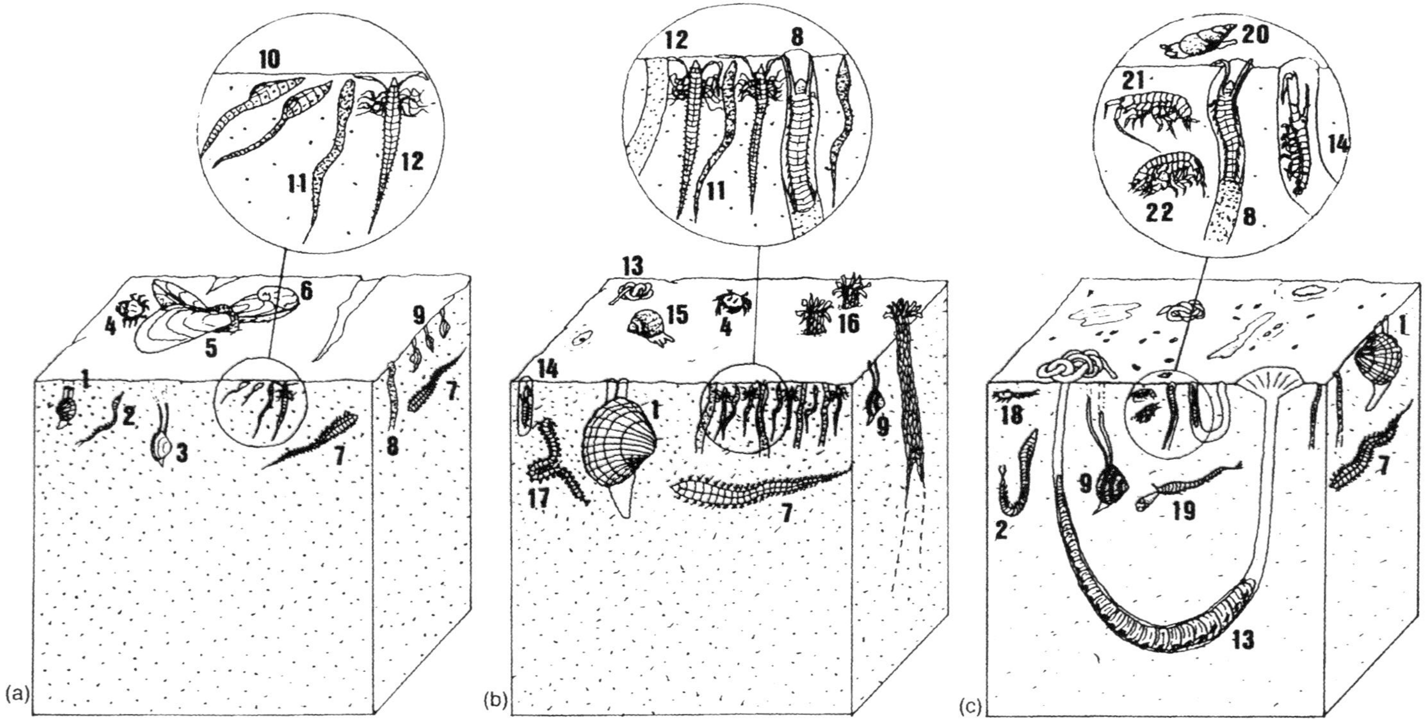

Figure 6.10 Stylized diagrams of the intertidal communities present at sites on the north shore of the outer Thames estuary: **(a)** Canvey Island; **(b)** Southend; **(c)** Shoeburyness East. **1** *Cerastoderma edule*; **2** *Scoloplos armiger*; **3** *Scrobicularia plana*; **4** *Carcinus maenas*; **5** *Mytilus edulis*; **6** *Crepidula fornicata*; **7** *Nephtys hombergi*; **8** *Pygospio elegans*; **9** *Macoma balthica*; **10** *Monopylephorus rubroniveus*; **11** *Tubificoides benedeni*; **12** *Caulleriella* spp.; **13** *Arenicola marina*; **14** *Corophium arenarium*; **15** *Littorina littorea*; **16** *Lanice conchilega*; **17** *Eteone longa*; **18** *Cumopsis goodsiri*; **19** *Glycera tridactyla*; **20** *Hydrobia ulvae*; **21** *Bathyporeia* sp.; **22** *Urothoë poseidonis*.

communities present at site from Canvey, Southend and Shoeburyness East, indicating the changes in the fauna associated with both the increase in salinity and the mean sediment particle size.

Total abundance of macroinvertebrates at Canvey tended to be an order of magnitude lower than the communities in the outer reaches of the estuary, the maximum total abundance recorded being $785/m^2$, although all sites were generally dominated by annelids. At Canvey the most common species were *Nephtys hombergi*, *Monopylephorus rubroniveus* and *Caulleriella* spp., though numbers of the latter two small species tended to be variable. *Nephtys* was usually the biomass dominant, but this depended on the presence or absence of large bivalve species such as *Cerastoderma edule* or *Mytilus edulis*. Most individuals present at Canvey tended to be small, particularly *Macoma balthica* and *Scrobicularia plana*. Other core species regularly recorded at Canvey included the polychaetes *Pygospio elegans* and *Scoloplos armiger* and the oligochaete *Tubificoides benedeni*, although never in the high abundances present at other intertidal sites. The core species at Canvey were irregularly supplemented by a large suite of other species, usually small individuals. These included 14 crustaceans, such as *Carcinus maenas*, *Corophium volutator*, *Gammarus salinus* and *G. locusta*, as well as more unusual species for an intertidal site such as *Atylus guttatus*, *Sphaeroma monodi*, *Corophium insidiosum* and *Caprella linearis*. Canvey was the only site where two species of crustacean were recorded: the ostracod *Sarsiella zostericola* and the praniza of a gnathiid isopod (probably *Paragnathia formica*). Young settling bivalves were also occasionally present in samples from Canvey, including *Mactra stultorum* and an *Ensis* sp. (?*arcuatus*).

The intertidal flats off Southend extend 2 km from the high water mark and appear highly stable, due to the increased sand content and the presence of many patches of sandmason worms (*Lanice conchilega*) and mussels (*Mytilus edulis*). The major disruption to the sediment appears to arise from the intensive bait-digging for lugworm (*Arenicola marina*) and king ragworm (*Nereis virens*) that occurs across large areas of the Southend flats. During 1989 and 1990, this site was characterized by large numbers of the small annelids *Tubificoides benedeni* ($1700/m^2$) and particularly cirratulids ($> 6000/m^2$), these being mainly *Caulleriella* spp. (*C. caput-esocis* and *C. zetlandica*) together with a lower proportion of *Tharyx marioni*. The numbers of these species decreased during 1991, resulting in a general decrease in the total abundance for the site, but this coincided with an increase in *Pygospio elegans* numbers (maximum recorded $400/m^2$). Total biomass for the site was not significantly affected, due to the presence of several large species – particularly *Cerastoderma edule*, *Macoma balthica* and *Nephtys hombergi*. *Nereis diversicolor*, when present, tended to be small. A change in *Corophium* species was notable between Canvey and Southend, from *C. volutator* to *C. arenarium*, presumably due to the increasing sand content preferred by

C. arenarium (Lincoln, 1979). Other core species regularly encountered at Southend included the polychaetes *Scoloplos armiger*, *Eteone longa* (up to 130/m^2) and a small capitellid species (?*Capitomastus giardi*), the shore crab *Carcinus maenas* (usually juveniles) and the oligochaete *Monopylephorus rubroniveus* (Figure 6.10b). The gastropod *Hydrobia ulvae* was uncommon at Southend when compared with the vast numbers often present on the sand flats further towards the open sea. Several species more associated with subtidal areas were recorded at Southend, including the polychaetes *Eulalia bilineata*, *Lepidonotus squamatus* and *Neoamphitrite figulus*, the amphipod *Microprotopus maculatus* and the bivalves *Nucula turgida* and *Fabulina fabula*. An interesting occurrence during the last quarter of 1990 was the leech *Theromyzon tessulatum*. This species can inhabit the nasal passages of wildfowl (Elliot and Mann, 1979) and this is the most likely method of transport from the freshwater reaches of the Thames and its tributaries.

The vast expanses of sand flats leading from Shoeburyness northwards along the Essex coast support a stable, high abundance, high biomass community comprising core species from a wide range of taxonomic groups (Figure 6.10c). During most of the year, the dominant species in terms of abundance, and often biomass, was the polychaete *Scoloplos armiger*, the maximum recorded density being 3460/m^2 (quarter 3/89) with a biomass of 45 g WW/m^2. At the end of each year large numbers of the gastropod *Hydrobia ulvae* were present in the samples, with densities up to 19 000/m^2. The main contributors to the biomass, in addition to *S. armiger*, were the bivalves *Macoma balthica* and *Cerastoderma edule*, which recorded densities of 140/m^2 at the Shoeburyness East site, together with the polychaete *Nephtys hombergi*. The composition of the smaller members of the macrofauna showed significant differences from sites further up the estuary. The small annelids *Tubificoides benedeni*, *Caulleriella* spp. and *Pygospio elegans* were generally present but in low numbers, the most common small organisms (apart from *H. ulvae*) in these outer sand flats being amphipods such as *Bathyporeia* spp., *Corophium arenarium* and *Urothoë poseidonis*. These peaked on 19 June 1991, when over 540 *Bathyporeia sarsi*/m^2 were present. Several other species were regularly recorded at the site, such as the polychaetes *Eteone longa*, *Anaitides mucosa* and *Glycera tridactyla* and the cumacean *Cumopsis goodsiri*. Due to its comparatively marine location and sandy substrate, the site at Shoeburyness East recorded several species not occurring in samples at other intertidal sites. These included three nemertean species (including *Cephalothrix rufifrons*), the polychaete *Nephtys cirrosa*, the amphipods *Bathyporeia pelagica* and *B. pilosa* and the bivalve *Mysella bidentata*. A further interesting occurrence was the opisthobranch *Retusa obtusa*, a predator on *Hydrobia* (Thompson and Brown, 1976) whose appearances coincided with the peaks in numbers of the small gastropod.

(b) A transect from Canvey to Allhallows

By undertaking a loose transect across the bed of the Thames between the north and south banks of the outer estuary, the main macroinvertebrate communities present in the rest of Zone 4 can be covered.

The mud banks present at Canvey are accompanied by thin areas of hard substratum running from mid-tide level towards the low water mark. These provide a point of attachment for beds of mussels, *Mytilus edulis*, on which grow large numbers of the American slipper limpet *Crepidula fornicata*. In the shallow subtidal reaches just off Canvey Island, the beds of *C. fornicata* become more extensive, the molluscs covering mussels, stones and other patches of hard substrata available. The resulting community recorded the highest total biomass of any assemblage in the Thames estuary (Table 6.1), mainly due to over 800 g WW of *Crepidula* tissue/m^2. The beds of slipper limpets provide a rare commodity in the Thames estuary – an area of hard substrate, which is colonized by species of bryozoan, hydroids, barnacles and sea anemones as well as providing a base for large mobile organisms such as the crabs *Carcinus maenas* and *Liocarcinus holsatus*, the starfish *Asterias rubens*, the sea mouse *Aphrodite aculeata* and the sea urchin *Psammechinus miliaris*. The mussels and slipper limpets are embedded in mud, giving a mixture of invertebrates from hard and soft substrata. Particularly common here were the polychaetes *Ampharete acutifrons* (240/m^2), *Eulalia bilineata* (98/m^2) and *Neoamphitrite figulus* (> 30/m^2) in addition to species abundant in the intertidal zone, such as *Scoloplos armiger*, *Nephtys hombergi* and *Nereis diversicolor*. The small annelids characteristic of the intertidal zone on the north shore were only present in small numbers, if at all.

Further south of Canvey lies the shipping channel. The depth of water increases very rapidly to a maximum of over 25 m. At the bottom of the shipping channel, below a marker called Chapman Buoy, there is a highly heterogeneous environment comprising muddy sand, clay, stones, shells and even sodden wood. The resulting macroinvertebrate community is by far the most species-rich and diverse in the whole estuary (Figure 6.11). The site consistently recorded the highest species number during the TEBP, a feat repeated for the rich meiofauna community recorded here (Attrill *et al.*, 1996b). On 14 October 1992 a total of 51 macroinvertebrates was recorded from four Day grabs, giving a cumulative total of over 100 species.

The site consistently recorded a high biomass (Table 6.1), due to the presence of many large species, often in high numbers. The most important of these include the sea anemone *Sagartia troglodytes* (maximum abundance of 360/m^2 and biomass of 115 g WW/m^2), the fan worm *Sabella pavonina* (265/m^2 and 90 g WW/m^2) and *Carcinus maenas* (30/m^2 and 71 g WW/m^2). Other smaller species, abundant and regularly occurring, that can also be regarded as part of the core community here include the

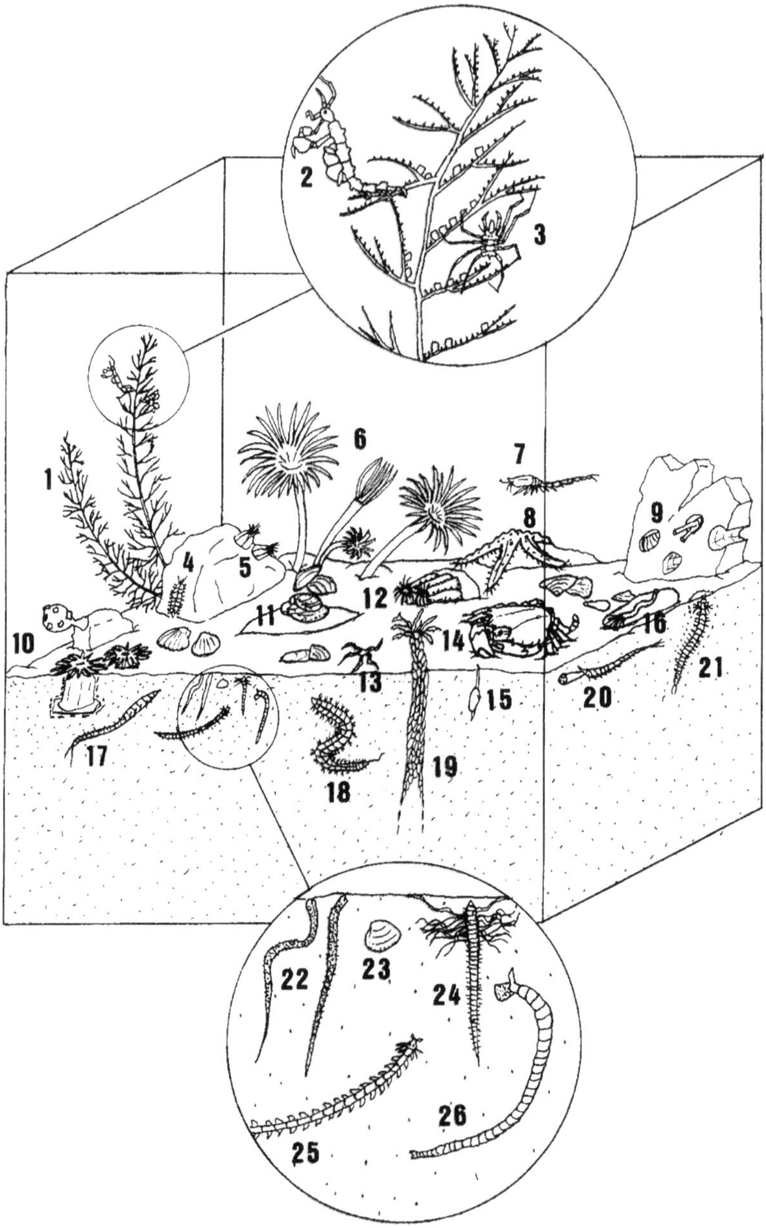

Figure 6.11 Stylized diagram of the community present at the base of the shipping channel beneath Chapman Buoy. **1** *Sertularia cupressina*; **2** *Caprella linearis*; **3** *Nymphon rubrum*; **4** *Lepidonotus squamatus*; **5** *Balanus improvisus*; **6** *Sabella pavonina*; **7** *Bodotria scorpioides*; **8** *Asterias rubens*; **9** *Barnea candida* and *Petricola pholadiformis*; **10** *Sagartia troglodytes*; **11** *Crepidula fornicata*; **12** *Sabellaria spinulosa*; **13** *Amphipholis squamata*; **14** *Carcinus maenas*; **15** *Abra alba*; **16** *Pomatoceros triqueter*; **17** *Scoloplos armiger*; **18** *Nephtys hombergi*; **19** *Lanice conchilega*; **20** *Glycera tridactyla*; **21** *Nereis longissima*; **22** *Tubificoides benedeni*; **23** *Nucula turgida*; **24** *Caulleriella* sp.; **25** *Eteone longa*; **26** Capitellidae sp.

polychaetes *Sabellaria spinulosa* (which forms substantial 'reefs'), *Ampharete acutifrons*, *Lepidonotus squamatus*, *Scoloplos armiger*, *Nereis longissima* and Capitellidae spp., the oligochaete *Tubificoides benedeni*, the sea spiders *Nymphon rubrum* and *Pycnogonum littorale*, together with the bivalves *Nucula turgida* and *Abra alba*. The clay walls of the channel and sodden wood present at the site contained large numbers of two boring bivalves: the pholid *Barnea candida* and the introduced venerupid *Petricola pholadiformis*. All species recorded from the subtidal area off Canvey were present at Chapman Buoy, with the notable exceptions of the two common intertidal species *Nereis diversicolor* and *Cerastoderma edule*. However, the community in the shipping channel was supplemented by a wide range of other species, several only recorded from this site. A total of 40 polychaete species have been noted, including scaleworms (*Gattyana cirrhosa*, *Pholoë synophthalmica*, *Harmathoë impar*, *Sthenelais boa*), paddleworms (*Mysta picta*, *Eteone longa*, *Anaitides mucosa*, *A. maculata*, *Eumida sanguinea*) and syllids (*Autolytus* sp., *Syllis gracilis*). Over 20 crustaceans were recorded (though their distribution was patchy), including the spider crabs *Hyas arenarius* and *Macropodia rostrata*, the amphipods *Corophium acherusicum* and *Atylus guttatus*, the prawn *Hippolyte varians* and the cumacean *Bodotria scorpioides*. Two brittle stars were regularly encountered: *Amphipholis squamata* and *Ophiura ophiura*. Other species uniquely recorded at Chapman Buoy included the sipunculid *Golfingia ?margaritaceum* and the sea spider *Achelia echinata*.

The great diversity of the site at the base of the shipping channel, together with the presence of many filter feeders and large species, suggests that this part of the estuary is highly stable in terms of the sediment dynamics and of the quality of water surrounding the site. It is possible that, due to its depth, the site does not come into contact with Thames water coming down the estuary to such a degree as shallower sites, the estuarine water floating on a wedge of heavier marine water.

The shallow subtidal area to the south of the shipping channel has a community that is markedly different in terms of richness and standing biomass from that of the area to the north. The substrate here is a fairly mobile muddy sand, the proportion of sand increasing with salinity. The fauna associated with this area (sites such as Blythe Sands and Grain Flats) is very poor when compared with the intertidal and other subtidal areas in the outer estuary. The community is characterized by low numbers of the more common species present at the other sites, with very little continuity from one sample to the next. The site at Blythe always recorded fewer than 10 species, the minimum being three. The only species to be present in all samples was the catworm *Nephtys hombergi*, the polychaete generally providing the main contribution towards the low biomass (maximum 8.34 g WW/m^2). Other regularly appearing species in this region of the Thames included the oligochaetes *Monopylephorus rubroniveus* and *Tubificoides benedeni*, small *Macoma balthica* and the cirratulid

Caulleriella spp. The remaining species recorded tended to be irregularly present and were generally the more mobile organisms such as Crustacea (*Corophium volutator, Crangon crangon, Diastylis bradyi,* mysids) or ephemeral polychaetes (*Pygospio elegans, Aricidia minuta, ?Capitomastus giardi*). The lowest total abundance recorded at Blythe was only 37.5/m^2 at the end of 1991, while the peak coincided with a group of *Bathyporeia sarsi* (380/m^2) – the only record of the species at this site.

The fauna seawards of Blythe at Grain Flats was very similar, characterized by low biomass and a variable community structure. Here too the dominant species was generally *Nephtys hombergi*, supplemented by the same range of core species with the addition of the cockle *Cerastoderma edule*. The number of species recorded overall was greater than at Blythe, probably due to the higher salinity regime, even though the maximum number of species recorded at any one occasion was only 12 (quarter 1/90). Additional species of interest included the amphipod *Atylus swammerdami*, the polychaetes *Anaitides mucosa* and *Goniada maculata*, the ophiuroid *Ophiura ophiura* and the molluscs *Nucula turgida, Hydrobia ulvae* and *Retusa obtusa*. The chaetognath *Sagitta elegans* was recorded in the benthos sample from the second quarter of 1989, which was unusual as this species is generally associated with oceanic waters.

The reasons for the depletion in the fauna across the shallow subtidal region of the southern outer estuary are not clear. However, the substrate at both sites, particularly Blythe, generally contained large amounts of decaying plant material not recorded in other parts of the estuary. Trawling across the area revealed large mats of decomposing seagrass (*Zostera*, Chapter 9). The nature of the currents in the outer estuary appears to aggregate detached seagrass along the southern coast of the outer estuary, which breaks up across the sediment. This material appears to have a detrimental effect on the fauna within the sediment, the species remaining tending to be large or mobile (e.g. *Nephtys, Macoma*). Mesocosm experiments at the Plymouth Marine Laboratories have indicated that applications of such plant material decreases the species richness of meiofauna in mud samples (Dr Mike Gee, personal communication). The meiofauna assemblages recorded at both Blythe Sands and Grain Flats tended to be comparatively poor.

The intertidal areas along the south of the outer Thames estuary extend up to 2 km from the high tide level and have a similar appearance to the mudflats off Southend. The sediment tends to be a sandy mud, with a slightly lower sand fraction than the large north shore areas discussed earlier, probably due to their more westwards location. The shore at Allhallows has three sections: a very narrow (10 m) sandy area by the seawall; an area of hard embedded stones and mussels (100 m); followed by the expanses of mud. Unlike Southend, there are very few patches of sandmason worms (*Lanice conchilega*) on the mud itself, or mussels (*Mytilus edulis*) away from the hard upper shore area.

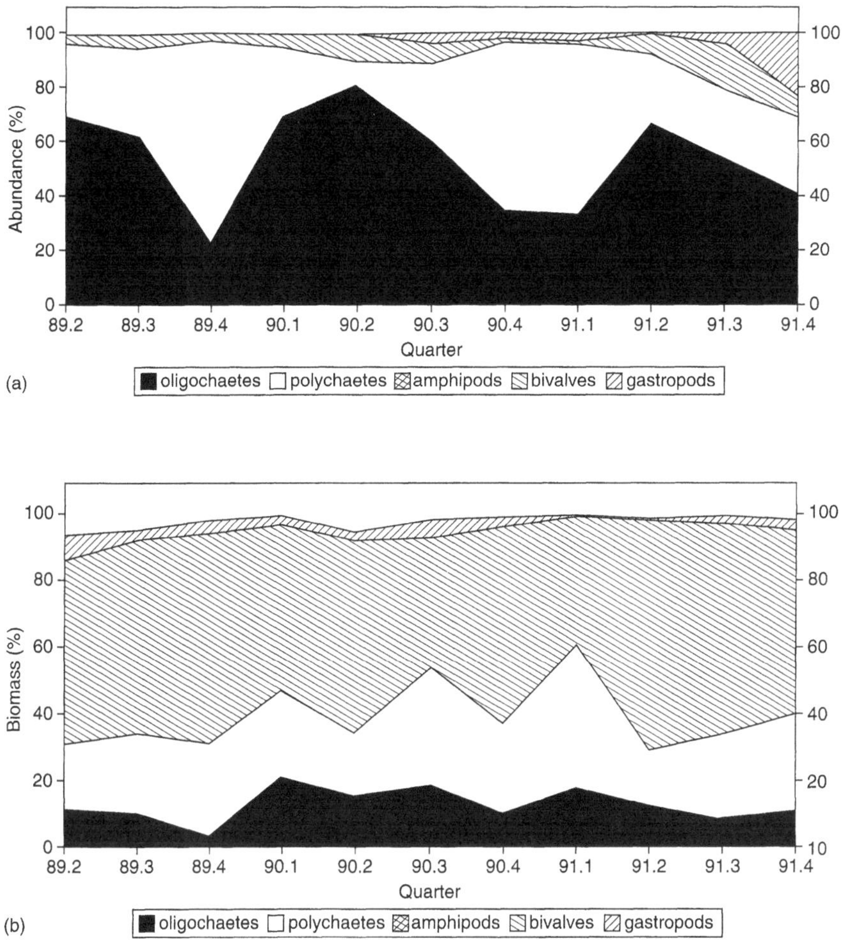

Figure 6.12 Percentage of (a) abundance and (b) biomass represented by each major group of organisms at Allhallows.

The mud is well compacted and stable, which appears to be reflected in the nature of the invertebrate community. The site has a higher number of species present in every sample than any other site, the proportions of which (in terms of biomass) remain remarkably stable (Figure 6.12b). The proportions of the abundance (Figure 6.12a) were influenced by variations in the numbers of small annelids present, particularly the oligochaete *Tubificoides benedeni* and the complex of cirratulid species (mainly *Caulleriella* spp.). Both were generally present in very high numbers, the maximum densities recorded being $4800/m^2$ for the cirratulids and $8200/m^2$ for *T. benedeni*, the highest density of this species recorded from the estuary. Over the period 1989–1991, the abundances of these small annelids were generally greater in the southern intertidal areas than the equivalent flats on the north shore of the estuary. The other core species constantly represented were the small capitellid (?*Capitomastus giardi*), *Nephtys hombergi*, *Corophium volutator* and particularly bivalve species *Scrobicularia plana* and *Macoma balthica*. Both bivalves recorded their highest densities at Allhallows: 214/m^2 for *S. plana* and 764/m^2 for *M. balthica*. Other core species only occasionally absent were *Carcinus maenas*, *Cerastoderma edule*, *Hydrobia ulvae*, *Eteone longa* and *Ampharete acutifrons*. Additional species recorded at Allhallows included many of the classic outer estuary intertidal species, such as *Mya arenaria*, *Littorina littorea*, *Pygospio elegans*, *Arenicola marina* and the isopod *Cyathura carinata*. More unusual occurrences were the marine centipede *Strigamia maritima* and the masked crab *Corystes cassivelaunus*. A notable absence from the intertidal area off Allhallows was the polychaete *Scoloplos armiger*. This was one of the most regularly occurring species at all the other sites in the outer estuary, particularly the intertidal areas on the north shore, but has not been recorded at Allhallows.

6.4.5 Zone 5: Outer Estuary II

From the ordination of sites using data from the second quarter of 1989 (Figure 6.2), Zone 5 comprises the two most seaward subtidal sites in the estuary, i.e. the area seawards of Southend. However, due the variable nature of the heterogeneous community at Chapman Buoy, this site was often placed in Zone 5 in subsequent surveys. This outer, subtidal zone can therefore be considered to cover the comparatively deep mid-channel area between Chapman Buoy and the outer limit of the Thames estuary (arbitrarily defined as discussed in Chapter 2).

The substrate in Zone 5 becomes progressively more uniform from Chapman Buoy seawards, the sediment at the outer site beneath the Sea Reach No. 2 marker buoy being a relatively clean, compacted sand. The site at Southend was located near the end of the long-sea sewage outfall, the substrate here being generally sandy with patches of harder material, such as shells and lumps of granite. This incongruous rock (granite not being a typical rock formation in south-east England!) was used to bury the extension to the long-sea outfall,

Large areas of Zone 5 are covered with beds of whiteweed – the hydroid *Sertularia compressina*. These beds are exploited by local fishermen, the hydroids being sold for decorative purposes, often to the United States (where it is dyed and used to decorate coffins). The *Sertularia* has its own epifaunal community, including delicate species such as amphipods and sea spiders, that are a contributing factor to the separation of the sites in Zone 5 from the rest of the estuary. The sandy substrate is firmly packed and comparatively stable, resulting in a high diversity community dominated by errant polychaetes, bivalves and amphipods. Oligochaetes were uncommon in Zone 5.

The biomass dominant in Zone 5 tended to be *Nephtys* spp., with *N. hombergi* prevalent at Southend and a complex of additional species (*N. caeca, N. cirrosa, N. longosetosa*) recorded at Sea Reach 2 (SR2). Other large core species present included the bivalve *Fabulina fabula* and the polychaetes *Scoloplos armiger* and *Glycera tridactyla*. However, the great majority of the species recorded from Zone 5 were small, resulting in comparatively low total biomass values (excluding *Sertularia* colonies). No one species dominated in terms of abundance, the most common species in the sample often depending on the number of *Sertularia* colonies entrapped in the grabs. These species included the polychaetes *Aricidia minuta* (maximum abundance 735/m^2 at SR2), which appears to fill the oligochaete niche in Zone 5, and *Scoloplos armiger* (128/m^2, SR2), the amphipods *Caprella linearis* (257/m^2, SEs), *Bathyporeia elegans* (250/m^2, SR2) and *Microprotopus maculatus* (83/m^2, SEs). *Nephtys hombergi* (170/m^2) and *Fabulina fabula* (105/m^2) were also present in significant numbers at Southend.

Several other species were regularly present in Zone 5, though in lower numbers. These minor core species covered a range of groups such as the polychaetes *Magelona mirabilis, Pygospio elegans* and *Caulleriella* sp., amphipods *Atylus falcatus* (Figure 6.13), *Periculoides longimanus*, the cumacean *Diastylis bradyi*, the sea spider *Nymphon rubrum* and the bivalves *Abra alba, Nucula turgida* and *Macoma balthica*. Zone 5 exhibited the widest range of crustacean species in the estuary, with a total of 40 species being recorded – 29 from the grab samples at Southend. Due to the saline nature of the Zone, many were unique to this area of the Thames, including the amphipods *Bathyporeia guilliamsoniana, Perioculodes longimanus, Panoploea minuta, Monoculodes carinatus* and *Pontocrates altamarinus*, cumacean *Pseudocuma longicornis*, isopod *Idotea linearis* and a juvenile of the crab *Atelecyclus rotundus*. Other interesting species encountered in the Zone 5 samples included the beautiful paddleworm *Anaitides groenlandica* (at SR2), the anemone *Metridium senile* (SR2), the spionid polychaetes *Spiophanes bombyx, Malacoceros fuliginosus* and *M. tetracerus*, several syllid polychaetes, including *Autolytus prolifera* (SEs), the heart urchins *Echinocardium cordatum* and *Echinocyamus pusillus* and small settled juveniles of a *Modiolus* species (*?phaseolinus*) that were recorded from the pieces of granite occurring in the Southend samples.

The outer estuary as a whole recorded several species as juveniles that appear not to survive to adult stages, these settling juveniles being frequently encountered in Zone 5. In addition to the *Modiolus* sp. and the *Atelecyclus* detailed above, young stages of *Ensis ?arcuatus, Corystes cassivelaunus, Mya arenaria* and *Mactra stultorum* were recorded. Young spider crabs (*Hyas arenarius, Macropodia rostrata*) were regularly present in grabs from the mid-channel area of the outer estuary, but adults were also frequently taken during trawling exercises.

Figure 6.13 *Atylus falcatus.*

An interesting feature of Zone 5 was the rich diversity of life to be found at the Southend site, situated at the end of the sewage outfall. The site recorded a total of 40 species from the four grabs taken at the end of 1991, second only to the extremely diverse site at Chapman Buoy. The fauna at the site shows no signs of influence from the outfall, the community being largely composed of delicate species generally intolerant to organic pollution. It is likely that the diffuser system on the end of the outfall is channelling the effluent vertically to disperse across the water surface, so bypassing the benthos near the pipe. The discharge would then settle over a wide area, depending on the tides, including the area around Sea Reach No. 2 Buoy. The community here, perhaps somewhat surprisingly, tended to record fewer species and was more variable than sites further into the estuary, possibly influenced by fine deposition of effluent.

6.5 CONCLUSIONS

Analysis of the benthic macroinvertebrate community by dividing the Thames estuary up into zones is useful in terms of describing the communities present. By the nature of the classification process, the areas of the Thames within each zone have similar faunal structures, allowing core species within that community to be defined. However, the zonation has further uses, both by providing a baseline structure allowing comparison with any future deviation from the patterns observed, and as a management tool in tandem with methods described in Chapter 2.

The Thames certainly now supports a rich assemblage of benthic macroinvertebrates over much of its length, which is a vast improvement on the situation present 20 years ago. It is likely that the estuary has reached a steady state in terms of its rehabilitation – a feature that may be confirmed with continued monitoring. The majority of possible improvements to polluting influences in the estuary have now been completed, though three factors remain that appear to be having detrimental effects of the community of invertebrates living within the estuary: namely, periods of low freshwater flow, the effect of the major sewage treatment works outfalls on neighbouring sites and periodic discharges of untreated sewage from the storm drains in the London area. Any scheme to alleviate the latter problem would involve vast cost, while a balance has probably been achieved between the needs to dispose of sewage effluent and 'acceptable' environmental damage to the mid-estuary. Therefore, as long as the quality of discharges and quantity of freshwater into the estuary is maintained, there is unlikely to be any further major improvement or deterioration in the benthic macroinvertebrate communities of the Thames estuary over the foreseeable future.

CHAPTER 7

Temporal changes in the movements and abundance of Thames estuary fish populations

Myles Thomas

7.1 INTRODUCTION

Estuaries play a crucial part in the life cycle of many fish species (Haedrich, 1983) and act as an important nursery area for many marine fish by providing a rich food source and protection from predation (Blaber and Blaber, 1980). The abrupt changes in oxygen concentration, temperature, turbidity, salinity and a number of other factors place considerable physiological demands on an organism. Adaptation to these factors is of course possible (Vernberg and Vernberg, 1976); however, the broad tolerance required to be successful within an estuary limits the number of niches available. Hence, the number of fish species that are truly estuarine is not great. Of the 112 species recorded in the tidal Thames (Appendix B), more than 90 are predominantly marine and only a small number, perhaps 10%, can be regarded as truly estuarine dependent.

The importance of the Thames estuary ecosystem can be shown by the value laid upon its fishery. The Ministry of Agriculture, Fisheries and Food (MAFF), for example, now regard the Thames estuary as probably the most important estuary on the English east coast for commercial flatfish (D. Eaton, MAFF, personal communication). However, this has not always been the case, as has been well documented (Wheeler, 1969a, 1979; Chapter 2).

The water quality of the tidal Thames was so poor from 1920 to 1964 that fish were absent from Fulham downriver to Tilbury during this

A Rehabilitated Estuarine Ecosystem. Edited by Martin J. Attrill.
Published in 1998 by Kluwer Academic Publishers, London. ISBN 0 412 49680 1.

period. The progressive recovery of the fauna from 1964 to 1973 has been reported elsewhere (Wheeler, 1969, 1979; Huddart and Arthur, 1971a,b,c; Sedgwick and Arthur, 1979) and highlighted the initial improvements in water quality after capital investment in the main sewage treatment works of the estuary. A second phase of improvement from the early 1970s up to 1980 was reported by Andrews and Rickard (1980) and Andrews (1984). Much of this work was based on surveys of fish entrained on the cooling water intake screens at West Thurrock Power Station and, with the exception of 1984, Thames Water Authority continued regular surveys at West Thurrock up to 1989, when responsibility for the work was passed to the National Rivers Authority (NRA). The NRA maintained this programme until the closure of the power station, for commercial reasons, in 1993. The Environment Agency is now developing new methods to provide information on other areas of the estuary which may be more representative of the system as a whole.

This chapter will review the data on the fish communities as indicated by West Thurrock Power Station samples from 1974 to 1991, supplementing information on invertebrates provided by Attrill and Thomas (1996). Changes in the composition and strength of the fish populations from 1985 to 1991 will be examined more closely and a possible alternative trawling technique for collecting semi-quantitative data from the Thames will also be described.

7.2 SAMPLING METHOD

Regular sampling first began at West Thurrock Power Station in 1974, primarily to monitor the effects of the improvements to Crossness and Beckton Sewage Treatment Works on the flora and fauna of the estuary. The power station was well situated (see Figure 1.1), being approximately 10 km downstream of the major sewage treatment works in a well mixed part of the estuary (St Clements Reach) with a salinity range sufficient to support a wide range of fish species. Although a number of qualitative surveys have been carried out since, the data collected from West Thurrock has produced the most complete data set on the Thames estuary fish populations.

The station extracts up to 30 million gallons (*c.* 136 million litres) of water per hour direct from the Thames for cooling purposes. This water enters the station via large culverts, the external openings of which are well below the water level at low tide.

The estuarine entrance to the intake is situated at the end of the power station's pier. Vertical slats, about 0.5 m apart, cross the entrance to prevent large flotsam and jetsam entering. The intake leads to a series of large open wells (Figure 7.1) and, as the water is pumped from them, they are filled with water from the estuary. From the wells, the water passes through the

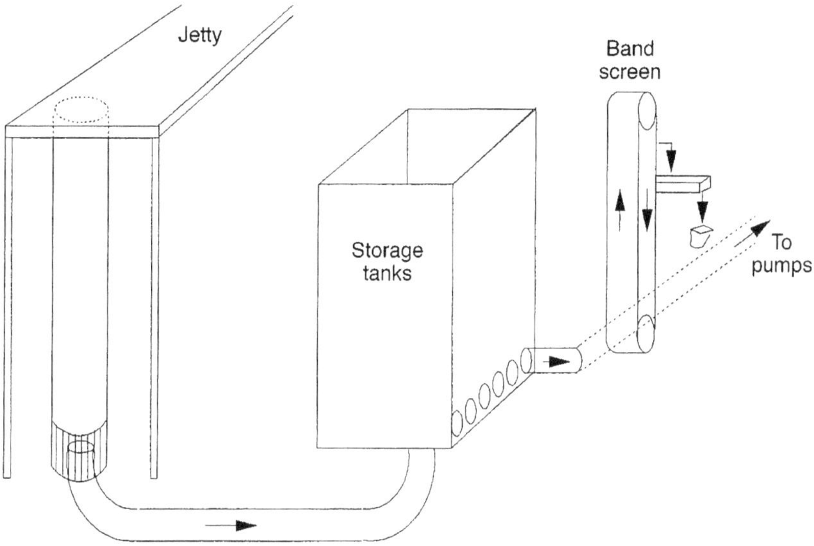

Figure 7.1 Cooling water intake system of West Thurrock Power Station.

upward-travelling section of a continually rotating band screen. The screens remove suspended solid matter < 1 cm in diameter from the water, since the final cooling pipes are only 2–3 cm in diameter. A variety of domestic objects and debris collects on the screen and, since the improvements in water quality, fish and other macrofauna (unexpected at the initial design phase) have become a major component.

At the top of the band, the rotating screen is washed clean by water jets and the filtered material, including fish, is flushed to a trash pit via a gutter/drain system. Fish caught in the trash pit can escape back to the estuary via a small outlet. The water used for cooling is finally returned to the estuary some distance upstream of the intake and is aerated as it is discharged.

Under normal conditions there were six screens in operation at West Thurrock, each having its own gutter to collect filtered material. For the purposes of the fish surveys, a square-framed bag with a mesh size of approximately 1 mm is fitted to the end of each gutter to intercept the screen washings.

All samples were taken on a spring tide commencing approximately half an hour before the morning low water and continued for four hours. Surveys during the 1970s often covered a six-hour period. To give consistency in the results, only data from the relevant four of the six hours is included here.

The sample nets were emptied every half hour during the survey period. After sorting, fish were identified to species following Wheeler

(1969b, 1992) and counted. The total lengths of all flatfish, gadoids, clupeids, bass (*Dicentrarchus labrax*) and smelt (*Osmerus eperlanus*) were measured to the nearest centimetre, and approximate numbers of invertebrates were recorded.

The trash pit was normally cleaned out early each day by National Power staff to prevent blockage of the outlet. Any species found here prior to sampling, but not found during the actual sampling period, were included in the species total for that sampling occasion as it is likely that these fish had only been in the system for a few hours.

The sampling technique has a number of advantages over other methods of obtaining fish samples. The method is less labour intensive than trawling, for example, and is also less susceptible to problems of gear damage and river traffic normally associated with a metropolitan estuary such as the Thames. In addition, semi-quantitative results are possible by determining the volume of water being drawn into the power station during the survey period.

The surveys carried out in the late 1970s suggested that the greatest diversity of fish species occurs on the flood tide. There is now some doubt as to whether this is still the case. A recent 24-hour survey indicated that the greatest diversity may occur on the ebb tide (Figure 7.2). Nevertheless, in order to continue the existing database in a comparable form, samples have been obtained following the sampling method first used in 1974.

The demand on the power station dictates the amount of cooling water needed and hence the number of intake pumps in operation (up to six). To account for this, and to allow comparisons to be made between surveys, a correction factor is applied to the results so that numbers of fish collected are expressed as the number expected if 100 million gallons (*c.* 455 Ml) had been sampled. This variation in abstraction rate will also vary the size of the area around the mouth of the intake pipe that is influenced by the suction effect of the intake. Hence, even with the correction factor applied, samples taken at times of significantly different abstraction rates are unlikely to be directly comparable.

7.3 SPECIES NUMBERS

Figure 7.3 indicates how the cumulative number of species found in the Thames between Fulham and Tilbury (a majority of which have been found at West Thurrock) has increased since the early 1960s. By the end of 1991 the total had reached 112 species and one hybrid, and consisted of 46 families. During the period 1964–1966, collection of fish was fairly random and relied largely on staff from various power stations reporting fish found on the cooling water screens. In 1967, the sampling effort increased and resulted in a significant increase in species numbers. Sampling effort has since remained fairly constant.

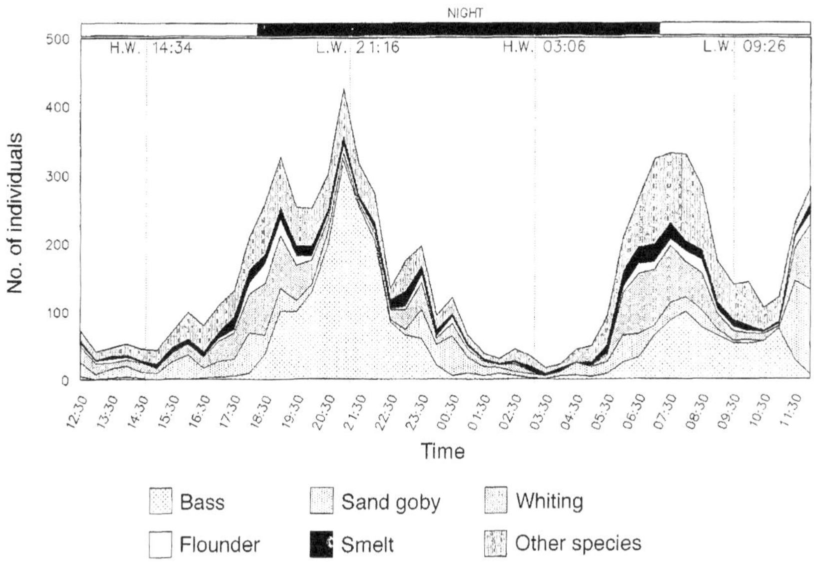

Figure 7.2 Fish abundance at West Thurrock Power Station during a 24-hour survey undertaken in October, 1989.

It should be emphasized that seasonal changes are not discernible in a graph of this type since any deterioration in species number would not be shown. Moreover, because the majority of species known to be present around the coast of south-east England have been recorded, further improvements will not be apparent.

The fish species present in any estuary can be categorized in a number of ways – for example: pelagic or demersal; or solitary or schooling. In the case of fish species appearing on the screens at West Thurrock the following categorization is most suitable:

- migratory (catadromous and anadromous), i.e. those species that move from freshwater to sea or from sea to freshwater for spawning;
- freshwater, i.e. those species that typically occur and breed in freshwater;
- estuarine, i.e. those species that spend a majority of their life cycle in or close to estuaries;
- marine estuarine-dependent, i.e. marine species that require an estuarine stage in their life cycle;
- marine straggler, i.e. marine species abundant in the marine environment but only infrequently found in estuaries.

Classification of the individual species of the Thames is not reproduced here as similar classifications can be found elsewhere (Claridge *et al.*, 1986; Elliott *et al.*, 1990).

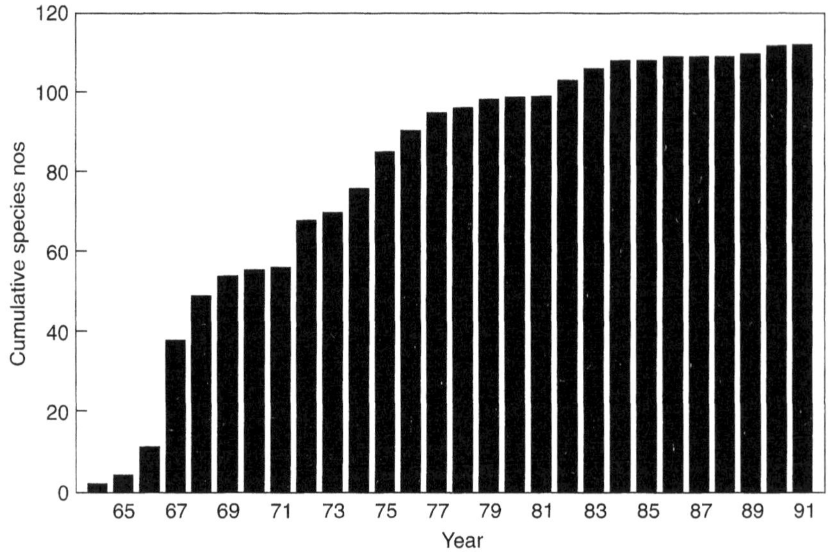

Figure 7.3 Cumulative fish species numbers, 1964–1991.

The seasonal movements of fish species associated with these categories determines the annual seasonal pattern in species numbers. This pattern is typified by the mean number of species for pooled fortnightly data over the 10-year period 1982–1991 (Figure 7.4). The seasonal trend of high species numbers in the autumn/winter and low numbers in the summer reflects the movement of marine species into and out of the estuary. Early in the year all categories of fish are present. During the spring the marine species begin to leave the estuary, so that by early summer only the catadromous, anadromous, estuarine and a few marine estuarine-dependent species remain. Marine stragglers and freshwater species are rarely found at this time of year. Later in the year, marine juveniles enter the river in large numbers and occasionally, during periods of high freshwater flow, these are accompanied on the intake screens by freshwater species.

Figure 7.5 shows the number of species per sample between 1974 and 1991. This is a better indication of the changes in species numbers than Figure 7.3. The general improvement over the nine years 1974 to 1982 is confirmed by the regression line for this period. In addition, the minimum number of species recorded each year increased from three in 1975 to 11 in 1982. For the period 1983–1991 no significant trends can be detected, though the range in annual species numbers is greater than the previous nine-year period. In the summer of 1986 a major storm over London caused the storm sewers to discharge into the Thames. The resulting breakdown of organic matter led to parts of the river in the central London area becoming anoxic. Numerous fish kills were reported and dissolved

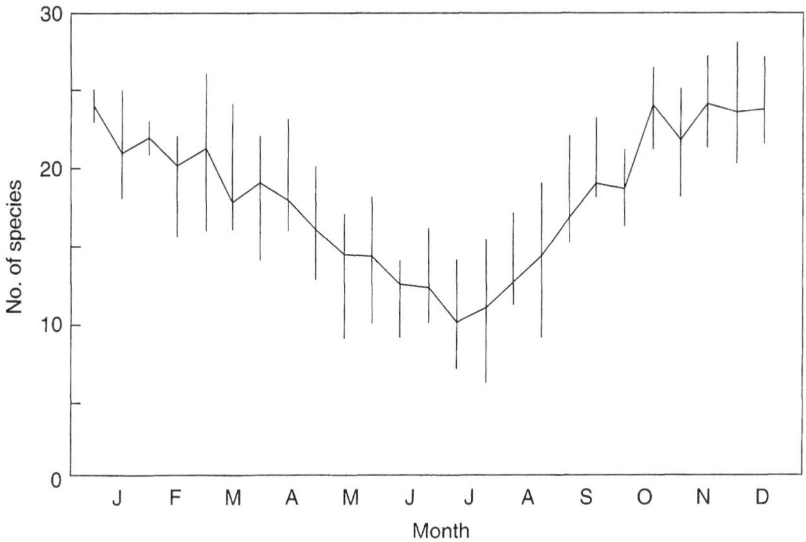

Figure 7.4 Mean number of species for pooled fortnightly data, 1982–1991.

oxygen levels as far downriver as West Thurrock dropped to a minimum of 15%. Species numbers reflected this water quality deterioration by dropping to just six during the worst period.

7.4 DIVERSITY

The Shannon–Wiener diversity index is perhaps the most popular measure of species diversity: it combines the measures of number of species and

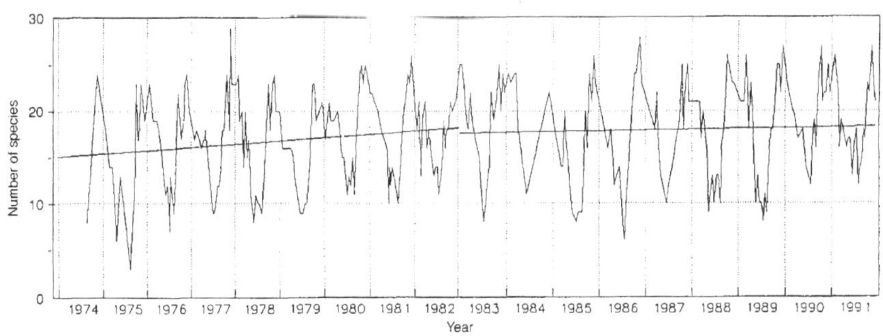

Figure 7.5 Number of species per sample at West Thurrock, 1974–1991, including the regression lines for the periods 1974–1982 and 1983–1991.

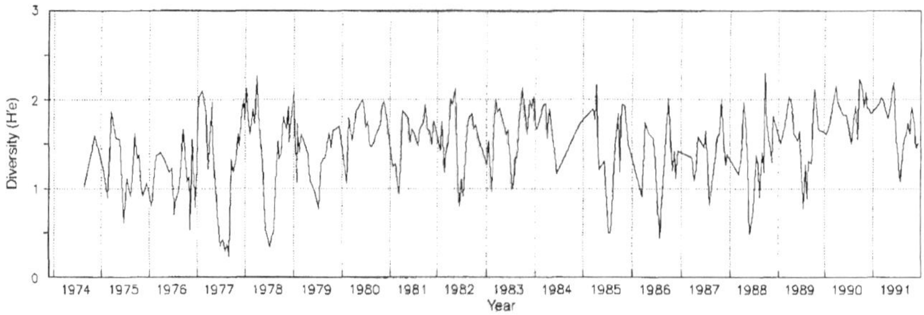

Figure 7.6 Diversity (H'e) per sample at West Thurrock, 1974–1991.

their relative abundance in the community. The species diversity for each sample between 1974 and 1991 is shown in Figure 7.6. Considerable variation in diversity can be seen over the 18-year sample period along with a similar seasonal effect shown by species number, indicating both high diversity in the autumn/winter and low diversity during the summer.

The diversity range of 0.24 to 2.3 over the 18-year sample period is similar to that recorded by Claridge *et al.* (1986) on the Severn estuary over a five-year period (0.8 to 2.3) and implies that the Thames, like the Severn, has a relatively more diverse fish fauna than other estuaries and that the use of cooling-water intake screens for fish sampling is particularly efficient.

Low diversity in the results reflects comparatively low numbers of species associated with the numerical dominance of a single species. In many communities this can be indicative of environmental stress; however, in 1977, 1978 and 1985 low diversity was due to the influx of large numbers of 0+ (first-year) flounder (*Platichthys flesus*) into the estuary. By comparison, in 1986 and 1988 low diversity was the result of a combination of exceptionally low species numbers, possibly related to environmental stress, and only moderately high numbers of 0+ flounder. Pronounced troughs in diversity can occur during the winter months when values are normally high. These are due to sudden increases in the number of whiting (*Merlangius merlangus*), sand goby (*Pomatoschistus minutus*), herring (*Clupea harengus*) and, in recent years, bass (*Dicentrarchus labrax*).

7.5 SPECIES ABUNDANCE

7.5.1 Flounder

Flounder (*Platichthys flesus*), along with the eel (*Anguilla anguilla*), was one of the first species to recolonize the Thames. These species are both able to tolerate reduced salinity and low dissolved oxygen levels. This is

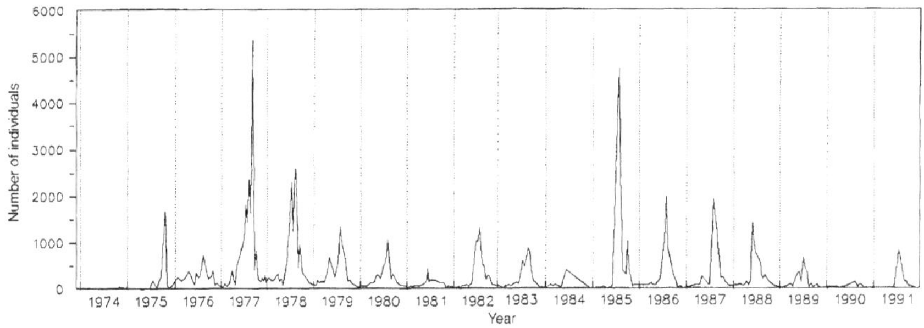

Figure 7.7 Number of flounder recorded in each sample from West Thurrock, 1974–1991.

illustrated in Figure 7.7, which also confirms the coincidence of high numbers of flounder with the low diversities.

Flounder are easily the most abundant of the Pleuronectidae in the Thames and spend the majority of their life cycle close to or within the estuary. An offshore migration precedes spawning in the spring, with the resulting 0+ juveniles entering the estuary in June/July. The juveniles penetrate well into the estuary and are normally recorded at the tidal limit at Teddington and in tidal creeks. They are also found in the freshwater tributaries where there is no restriction by physical barriers. Rapid growth follows the migration into the estuary with modal length increasing from 4 cm in June to 7 cm in December and around 10 cm by the following June.

In 1977 and 1985 (particularly good years for flounder), more than 4500 were recorded in single surveys in both years. A decline in flounder abundance has been apparent at West Thurrock since the mid-1980s, reflecting a reduction in the number of 0+ fish within the population.

Initial examination of this data might suggest that the reduction in juveniles could be linked with a detrimental change in the estuarine environment. However, it would appear from the length frequency data that the 0+ fish are not entering the estuary, suggesting that the reproductive success is poor and/or the ability of the fish to reach the metamorphosis stage (1.5–3 cm) is being affected by excessive mortality. This mortality has been attributed to predation on the drifting eggs and larvae (e.g. by Ctenophores; Kuipers *et al.*, 1990), or starvation of the larvae (Harding *et al.*, 1978). There is no evidence to suggest that reproduction is any less successful than previous years and predation is the most likely cause. Van der Veer (1985) has suggested that coelenterate predation can have a significant effect on the abundance of flatfish larvae in the Dutch Wadden Sea. In that case, the sharp increase in numbers of coelenterates coincided with a dramatic drop in 0+ flatfish numbers. Moreover, flounder larvae were regarded the most susceptible 'because of the shorter period of immigration and the coincidence between coelenterate outburst and

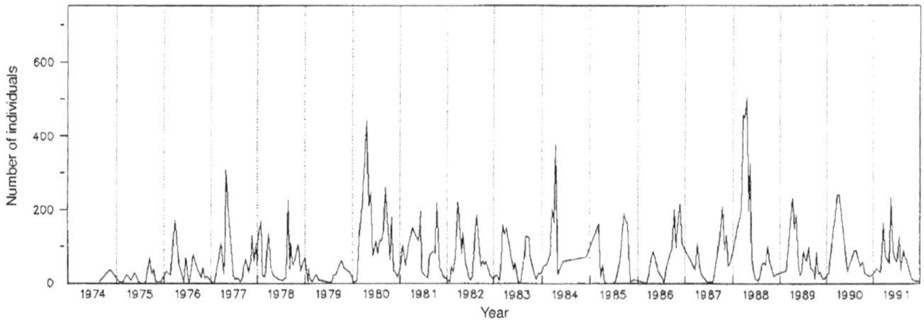

Figure 7.8 Number of sole recorded in each sample from West Thurrock, 1974–1991.

maximum flounder abundance'. The ctenophore *Pleurobrachia pileus* was found to predate almost entirely on flounder larvae. Qualitative data collected from the cooling water intake screens at West Thurrock suggests a significant increase in *P. pileus* since the mid-1980s (Attrill and Thomas, 1996), which coincides with the reduction in flounder abundance in the Thames estuary in recent years. Further work is needed to assess the effects of predation on the recruitment of 0+ flounder into the Thames.

7.5.2 Sole

Dover sole (*Solea solea*) is a marine estuarine-dependent flatfish species of commercial importance that can occur in such large numbers that the Thames can be regarded as one of the most important nursery grounds in England and Wales (D. Eaton, MAFF, personal communication).

Sole were relatively uncommon during the mid-1970s, due to relatively poor water quality. The number present in samples reached peaks of 400 and 500, respectively, in the springs of 1980 and 1988 (Figure 7.8). North Sea sole spawn from April to June (Whitehead *et al.*, 1989) but, although this species has been reported as spawning as far up the river as Gravesend (Riley *et al.*, 1981), 0+ fish do not occur on the screens at West Thurrock until late July and are typically 5–6 cm in length. These fish are the most abundant year class throughout the year. Sole, unlike most other common marine species in the estuary, migrate offshore from October to December (Wheeler, 1969b) and hence their abundance declines during the winter. The juveniles re-enter the river in early spring and numbers gradually reduce throughout the rest of the year.

7.5.3 Whiting

Whiting (*Merlangius merlangus*) is the most common gadoid in the Thames estuary and is predominantly represented by 0+ fish on the cooling water

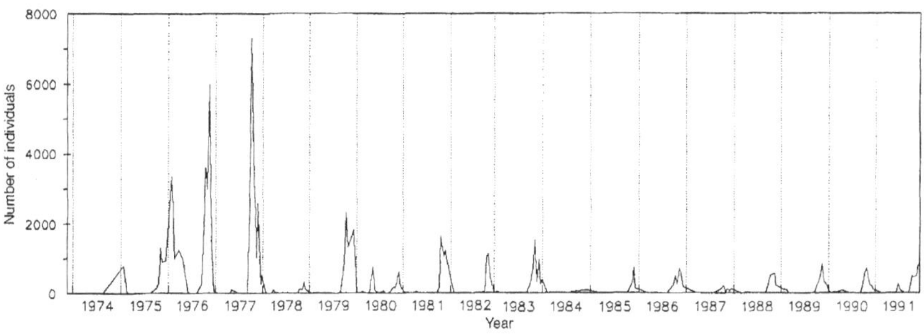

Figure 7.9 Number of whiting recorded in each sample from West Thurrock, 1974–1991.

screens. In the north-eastern Atlantic, spawning occurs between January and July (Whitehead *et al.*, 1989) but is likely to peak in the spring in the southern North Sea. Juveniles are fairly rare throughout most of the year but fish of 10–12 cm in length can appear in very large numbers in the autumn.

Whiting was the first marine species to recolonize the estuary in significant numbers and was present in very high numbers in the consecutive autumns of 1975, 1976 and 1977 (Figure 7.9). This contrasts with the abundance of around eight per sample in the autumn of 1971 (Wheeler, 1979). The reduced abundance since the mid-1970s is not necessarily a reflection of changes in the estuarine environment, but is likely to be due to variations in spawning success, increased competition from other species, environmental variations or fishing pressures in the North Sea.

7.5.4 Bass

Spawning success, due possibly to recent mild winters and warmer temperatures, is also likely to be the cause of a significant increase in juvenile bass (*Dicentrarchus labrax*) in 1989 (Figure 7.10). In British waters spawning occurs offshore in deep water from March to June and occasionally inshore, probably by first time spawners (Pickett, 1989). The post-larvae, approximately 1.5–3.0 cm in length, are thought to enter the estuary from June to August and do not appear on the screens until mid to late August, by which time they have reached approximately 4.0–5.0 cm.

Like most other marine species, bass are represented mainly by 0+ fish on the cooling water screens. However, the presence of second-year and a few third-year fish in recent years has indicated a different migratory pattern dependent on age: 0+ fish enter the estuary in late summer, followed two months later by lesser numbers of second- and third-year fish. The reverse occurs when, in the following spring, the older and then the younger fish move out to sea.

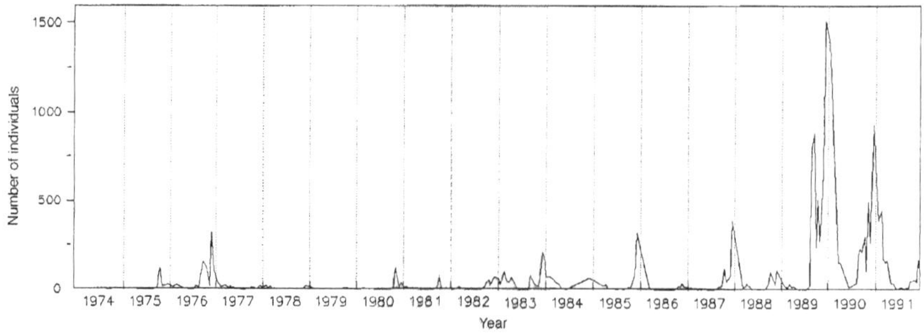

Figure 7.10 Number of bass recorded in each sample from West Thurrock, 1974–1991.

7.5.5 Sand goby

Currently the most abundant species found during the surveys is the sand goby (*Pomatoschistus minutus*). Of no commercial importance, the sand goby appears on the screens in very large numbers in the autumn. Since they live for only 1–2 years, the fish that are caught tend to be represented by a single age group. Spawning occurs inshore during the spring (Whitehead *et al.*, 1989) but is not thought to take place in estuaries (Miller, 1963). Data for the 18-year sample period appears to suggest a 10-year cycle in sand goby abundance, indicated by peaks in abundance in both 1976 and 1986 with a corresponding decline in the following years (Figure 7.11).

7.5.6 Herring

Although all members of the Clupeidae have been recorded in the Thames, the herring (*Clupea harengus*) is by far the most abundant. The

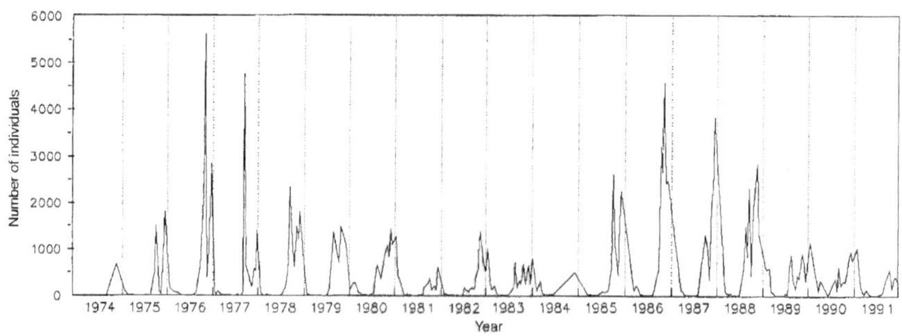

Figure 7.11 Number of sand gobies recorded in each sample from West Thurrock, 1974–1991.

commercial importance of this group is shown by its contribution of 20–25% of the total world fish catch with an annual value of £1 billion (Blaxter, 1990). Herring can be divided into numerous more or less distinct races which differ from one another both morphologically and biologically. The North Sea herring population comprises a small number of different races or breeding groups, each with their own well-defined spawning grounds. These different races spawn to the east of Orkney (July), off the east coast of Scotland (August and September), in the central North Sea (October) and in the southern North Sea and English Channel (November and December) (Blaxter, 1986; Cushing, 1986). The latter group is likely to be the main source of Thames herring, which appear on the screens in greatest numbers during January and February. These fish measure around 8 cm in length, and are about one year old (Wheeler, 1969b).

Herring did not appear in the Thames in significant numbers until the late 1970s (Figure 7.12), possibly due to a sensitivity to relatively poor water quality. The fluctuation in its numbers over the sample period is likely to be a reflection of the intense fishing pressures on this species and spawning success.

7.5.7 Smelt

Smelt (*Osmerus eperlanus*) were found in the estuary in such abundance prior to the water quality problems that it supported a commercial smelt fishery. As a relative of the salmon (*Salmo salar*), this species is regarded as being sensitive to water quality and did not reappear in the Thames in significant numbers until 1977 (Figure 7.13). Abundance has varied from year to year, with the greatest numbers occurring in 1979 and 1982. A general decline was indicated during the late 1980s: 1989 and 1991 were particularly poor years, with the former producing the lowest abundance since 1977.

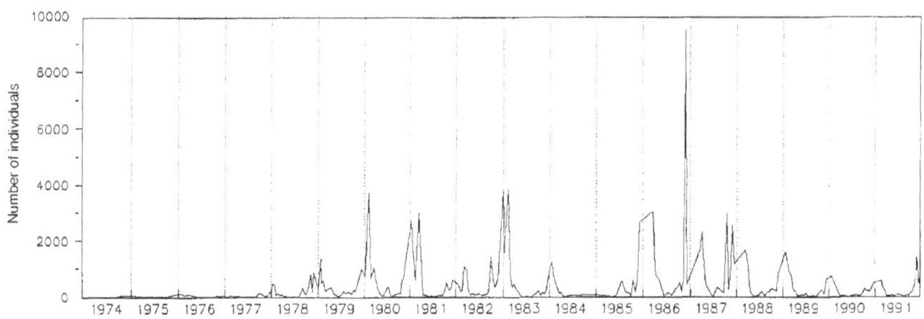

Figure 7.12 Number of herring recorded in each sample from West Thurrock, 1974–1991.

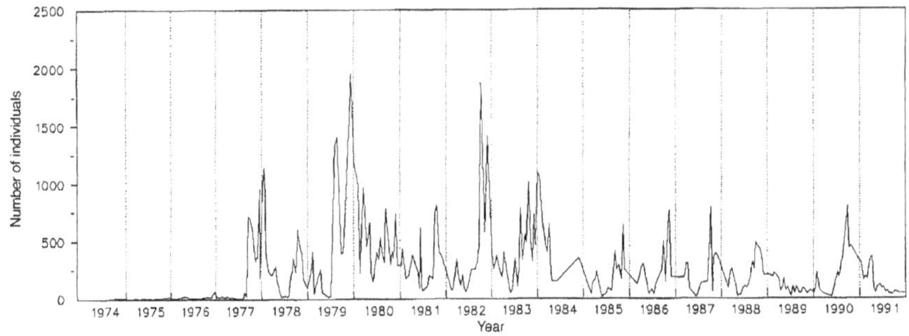

Figure 7.13 Number of smelt recorded in each sample from West Thurrock, 1974–1991.

In July, 0+ fish first appear on the screens, at a size of 6–8 cm, and they reach approximately 10 cm by December. Older fish tend to occur throughout the year in lesser numbers. Further information on the age of these fish is difficult to determine from the length data.

Smelt spend the majority of their life cycle within the estuary and, therefore, are likely to be good indicators of overall water quality. They spawn in the less saline reaches of the estuary, such as Chelsea and Wandsworth. These areas are susceptible to elevated salinity levels, exposure, and reduced flushing and dilution of pollutants due to recent low flows in south-east England. However, major variations in the abundance of smelt are known to occur naturally due primarily to disease (Hutchinson, 1983) and possibly parasitic infection (Chapter 8). Further study of populations within the Thames is necessary in order to determine whether environmental change is the cause of the recent reduction in abundance and to assess the value of this species for water quality monitoring purposes.

7.5.8 Salmon

The Atlantic salmon (*Salmo salar*) is uncommon in the Thames estuary, however its return to the river has received a considerable amount of publicity in recent years. The first salmon to be found in the Thames estuary for 140 years was found alive on the screens at West Thurrock in 1974. A Working Party on Thames Migratory Salmonids was set up in 1978 to consider the reintroduction of salmon to the river and concluded that the project would be both 'feasible' and 'worthwhile' (Anon., 1986). The Thames Salmon Rehabilitation Scheme, a 17-year project consisting initially of three phases, began in 1979 and by 1981 the first significant numbers of grilse had returned. The main objectives of the scheme were the stocking of parr, rearing of juveniles, capturing of ascending adults and modification of weirs to allow access to the upper reaches of the Thames catchment. Further studies on the source of stock fish and their survival

Table 7.1 Adult salmon returns to the Thames, 1970–1991

Year	*No. returned*	*Year*	*No. returned*
1970	0	1981	8
1971	0	1982	128
1972	0	1983	90
1973	0	1984	106
1974	1	1985	75
1975	2	1986	176
1976	1	1987	58
1977	0	1988	323
1978	1	1989	133
1979	1	1990	154
1980	4	1991	59

rate were undertaken during the first phase of the scheme, along with investigations into the ability of these fish to pass safely down through the catchment to the feeding grounds (Gough, 1990).

By 1986, significant numbers of grilse (Table 7.1) were being recorded, representing a run of 200 to 300 salmon. However, the runs of both 1986 and 1987 were affected, to varying degrees, by poor water quality in the tidal Thames, confirming that conditions were not ideal for an uninhibited return to the river. In recent years, water quantity (in terms of freshwater flow) rather than quality appears to have been the most significant factor.

7.6 COMMUNITY COMPOSITION

The Gadidae was the most diverse of the 46 families recorded at West Thurrock since 1964, represented by 15 species. This family is closely followed by the Cyprinidae with 13 species and the Gobiidae with eight species. The amount of data on numbers of individuals collected since 1974 is too extensive to summarize here; hence a seven-year period from 1985 to 1991 will be used to illustrate the changes in species abundance. During this period, the Gobiidae were the most abundant (66 599), overwhelmingly dominated by the sand goby (99.8%) (Table 7.2). Clupeidae were ranked second (63 563), represented by five species, with the four species of Pleuronectidae ranking this family third (38 678). The numbers of Osmeridae, represented by a single species (21 449), and the Gadoids (17 678), with 10 species, ranked these families fourth and fifth, respectively.

The top 10 most abundant species accounted for 95.3–99.5% of the total annual catch. The numbers of sand goby and herring made a far greater contribution to the total catch than any other species. Since the surveys were undertaken during what is considered to be the peak of species numbers and abundance, care should be taken in extrapolating the data over a 24-hour period.

Table 7.2 Species numbers (N), relative abundance (%) and rank (R) for the 20 most common species, 1985–1991

	1985			*1986*			*1987*			*1988*		
Species	*N*	*%*	*R*	*N*	*%*	*R*	*N*	*%*	*R*	*N*	*%*	*R*
Sand goby	8783	26.9	2	16356	33.5	2	10777	30.0	2	13154	32.7	1
Herring	5342	16.3	3	17307	35.4	1	13595	37.9	1	8816	21.9	2
Flounder	10994	33.6	1	5201	10.6	3	3597	10.0	3	6693	16.6	3
Smelt	2665	8.2	4	3423	7.0	4	3025	8.4	4	3590	8.9	4
Bass	535	1.6	9	86	0.2	16	700	2.0	6	470	1.2	8
Whiting	1175	3.6	5	2328	4.8	5	598	1.7	8	1756	4.4	6
Sole	862	2.6	7	1145	2.3	7	907	2.5	5	2579	6.4	5
Sprat	563	1.7	8	1446	3.0	6	698	1.9	7	437	1.1	10
Nilsson's pipefish	942	2.9	6	350	0.7	9	272	0.8	11	671	1.7	7
Eel	194	0.6	11	337	0.7	10	391	1.1	9	373	0.9	12
Pouting	37	0.1	16	131	0.3	15	21	0.1	19	137	0.3	15
Plaice	226	0.7	10	731	1.5	8	259	0.7	12	229	0.6	13
Pogge	30	0.1	17	323	0.7	11	243	0.7	13	455	1.1	9
Dab	63	0.2	12=	227	0.5	12	201	0.6	14	399	1.0	11
Three-spined stickleback	63	0.2	12=	134	0.3	14	282	0.8	10	166	0.4	14
Five-bearded rockling	6	< 0.1	22=	8	< 0.1	25	4	< 0.1	25=	8	< 0.1	24
Poor cod	51	0.2	15	61	0.1	17	51	0.1	16	68	0.2	17
Common sea snail	54	0.2	14	202	0.4	13	109	0.3	15	98	0.2	16
Thin-lipped mullet	3	< 0.1	25=	2	< 0.1	31=	0	0	–	3	< 0.1	29
Tub gurnard	8	< 0.1	21	18	< 0.1	21	12	< 0.1	20	16	< 0.1	20

Table 7.2 (*continued*)

	1989			*1990*			*1991*			*Total*		
Species	*N*	*%*	*R*	*N*	*%*	*R*	*N*	*%*	*R*	*N*	*%*	*R*
Sand goby	6624	21.7	1	6415	22.8	1	4343	17.8	2	66452	27.6	1
Herring	5141	16.8	3	2926	10.4	4	4504	18.5	1	57631	24.0	2
Flounder	3604	11.8	4	1303	4.6	6	2962	12.1	4	34534	14.3	3
Smelt	1904	6.2	6	4361	15.5	3	2481	10.2	5	21449	8.9	4
Bass	5470	17.9	2	5683	20.2	2	1903	7.8	6	14847	6.2	5
Whiting	2572	8.4	5	2504	8.9	5	3508	14.4	3	14441	6.0	6
Sole	1584	5.2	7	1145	4.1	7	1356	5.6	7	9578	4.0	7
Sprat	548	1.8	9	965	3.4	8	1269	5.2	8	5926	2.5	8
Nilsson's pipefish	1553	5.1	8	361	1.3	11	518	2.1	9	4667	1.9	9
Eel	315	1.0	11	225	0.8	12	220	0.9	12	2055	0.9	10
Pouting	113	0.4	15	742	2.6	10	435	1.8	10	1616	0.7	11
Plaice	43	0.1	19	9	< 0.1	27	64	0.3	15	1561	0.6	12
Pogge	320	1.0	10	95	0.3	16	32	0.1	19	1498	0.6	13
Dab	138	0.5	13	31	0.1	18=	25	0.1	21	1084	0.5	14
Three-spined stickleback	140	0.5	12	119	0.4	14	82	0.3	13	986	0.4	15
Five-bearded rockling	72	0.2	16	786	2.8	9	26	0.1	20	910	0.4	16
Poor cod	49	0.2	18	115	0.4	15	306	1.3	11	701	0.3	17
Common sea snail	25	< 0.1	20	4	< 0.1	28	1	< 0.1	33=	463	0.2	18
Thin-lipped mullet	125	0.4	14	140	0.5	13	72	0.3	14	339	0.1	19
Tub gurnard	20	< 0.1	22=	76	0.3	17	59	0.2	17	209	0.1	20

The annual relative abundance of each species ranked the sand goby first or second in all years; herring was always in the top four and flounder in the top six. Bass have shown the greatest change over the years, from a rank of 16th in 1986 to 2nd in both 1989 and 1990.

7.6.1 Species dendrogram

The classification dendrogram for the 69 species recorded at West Thurrock during the seven-year period 1985–1991 is shown in Figure 7.14. The dendrogram was constructed using a Bray–Curtis measure of similarity on double square root transformed data (to reduce the influence of abundant species) based on species abundance in each sample.

In general, the species showed no significant tendency to separate into well-defined groups. However, 16 of the 20 most common species did group together at the 52% level and included the three most common gadoids – whiting, pouting (*Trisopterus luscus*) and poor cod (*Trisopterus minutus*) – together with three of the four Pleuronectidae species: flounder, plaice (*Pleuronectes platessa*) and dab (*Limanda limanda*). In fact, even closer associations are indicated by two species within both of these groups; pouting and poor cod grouped together at the 67% level, while dab and plaice grouped at the 79% level.

Sole and flounder showed a high level of similarity (80%) reflecting the two species' low salinity tolerance and benthic preferences. A surprisingly high association between smelt and herring of 83% is also indicated, possibly due in part to the presence of both species in varying abundances throughout the year. The 100% similarity between four-bearded rockling (*Enchelyopus cimbrius*) and lemon sole (*Microstomus kitt*) is due to the presence of one individual of each species occurring in a single sample over the seven-year survey and illustrates how a high level of grouping involving uncommon species should be viewed with caution.

7.7 TRAWLING METHOD

Although capable of producing a good database on the state of the fish populations in this part of the river, there are disadvantages in using the results from a single source such as West Thurrock to provide representative assessments of the structure of the fish community in the Thames estuary as a whole. The fixed position can only give an indication of the fish communities in a small area. Moreover, fish species drawn into the power station may consist of the smaller, weaker fish that are not strong enough to avoid the suction effect of the intake and hence a bias of results may occur. This selectivity was confirmed by qualitative trawls undertaken by the then NRA adjacent to West Thurrock during simultaneous surveys of the screens. The results showed that, although in low

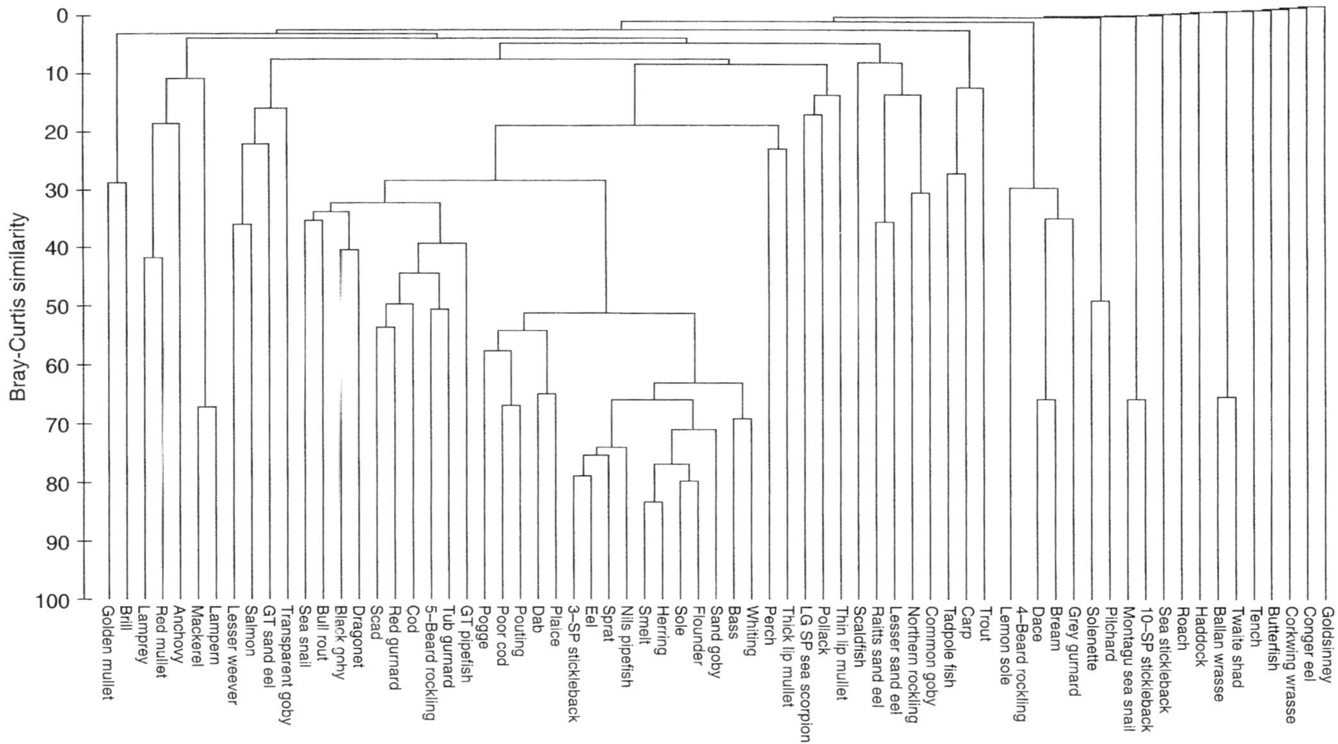

Figure 7.14 Dendrogram showing the classification of 69 fish species collected from West Thurrock based on the abundance of individuals in each sample from 1935–1991.

numbers, larger fish were present in the river than were appearing on the screens.

A technique was required that would overcome these problems and the practical difficulties associated with sampling a metropolitan estuary such as the Thames. In addition, at least semi-quantitative results were needed. Experimentation with a number of trawling techniques in the Thames estuary in early 1991 resulted in the selection of a Lowestoft Frame Trawl as the most suitable technique for collecting semi-quantitative fish data. The main advantage of this technique is that, by adjustment of the net's diving plane and/or the speed of the towing vessel, the net can be positioned at any level of the water column, thus reducing the bias that some techniques may have to specific areas of the water column. A number of adaptations were made to the net design and trawling method originally reported by Walker and Davies (1986). The net modifications included: use of a stronger net with a smaller stretched mesh size of 20 mm in the cod-end; adding a rope grid across the mouth of the frame to prevent large debris and boulders from entering; shortening of the bridles for greater stability and ease of emptying; the use of a General Oceanics 2030 Series Flowmeter; and the absence of a depth transducer.

A 12-month trial period was completed in August 1992 and it is hoped that this technique will continue in order to appraise its suitability for collecting representative data. A more comprehensive review of the work will be reported at a later date, but the first year's results are included here to complement the data from West Thurrock.

7.7.1 Method

Due to difficulties in finding a suitable vessel for the upper estuary, only sites below Tower Bridge were sampled. Each site was surveyed on four occasions (once per quarter) between September 1991 and August 1992. Sites were located using DECCA navigational equipment and, when possible, surveyed within two hours of low water. The commercial fishing trawler 'Ina-k' was used at all sites to deploy two nets for each 15-minute subsample. During the first subsample, the nets were positioned on the bed of the estuary and towed against the tide. The second subsample was collected by positioning the nets in mid-water and a further three subsamples (two on the bed and one in mid-water) were taken. Collection of all five subsamples required a total survey time of 75 minutes.

All fish were sorted and identified to species following Wheeler (1969b, 1992), counted and selected species were measured. Invertebrate species abundance was also estimated.

Flowmeter readings were recorded after every subsample, from which an estimate of the volume of water sampled and distance covered was made. The results were then standardized to 5 million gallons (*c.* 22.7 Ml).

7.7.2 Results

The results for each quarter are summarized in Figures 7.15 and 7.16. Spatial comparisons show West Shoebury to be poor in terms of fish diversity; however, the total number of species (17) recorded at this site over the four quarters is high and greater than the number of species recorded at both Blackwall and Barking reaches (11 and 12, respectively). Blythe Sands produced the most species in a single quarter (16 in the second quarter of 1992), whilst the greatest number of species recorded over the four quarters was at The Nore (22). Overall, 31 species were recorded at these six sites compared with the 42 species found in surveys at West Thurrock during 1991.

(a) Fourth quarter of 1991

Twenty-one species were recorded during this quarter, 15 of which were found at Blythe Sands.

The relative abundance of herring and whiting was greatest in the mid-estuary and highest at Blackwall and St. Clements. The presence of these marine species at these sites and at this time of year reflects their usual overwintering behaviour.

Sole, although fairly numerous at St Clements Reach, were generally scarce at all sites, confirming the early autumn migratory pattern of returning to the sea. Plaice have a much higher salinity preference compared with a number of other flatfish (Riley *et al.*, 1981) and were recorded no further into the estuary than Blyth Sands. Flounder, being more tolerant of low salinity, were present, along with smelt, at most of the upper sites (Blackwall, Barking and St Clements Reaches).

(b) First quarter of 1992

Of the 20 species caught during this quarter, 15 were recorded at St Clement's Reach.

The numbers of herring and whiting began to decline as the juveniles began their seasonal migration back to the sea. The abundance of both sole and flounder was unchanged, although the relative abundance of flounder at Blackwall Reach was a little higher. Plaice remained present at the three outer sites. Smelt were more common in the upper sites, reflecting the spawning run to the freshwater spawning areas in the upper river.

(c) Second quarter of 1992

This quarter produced both the highest species number during the survey period (24) and the greatest range in the number of species, with Blyth Sands recording 16 species and West Shoebury recording just four.

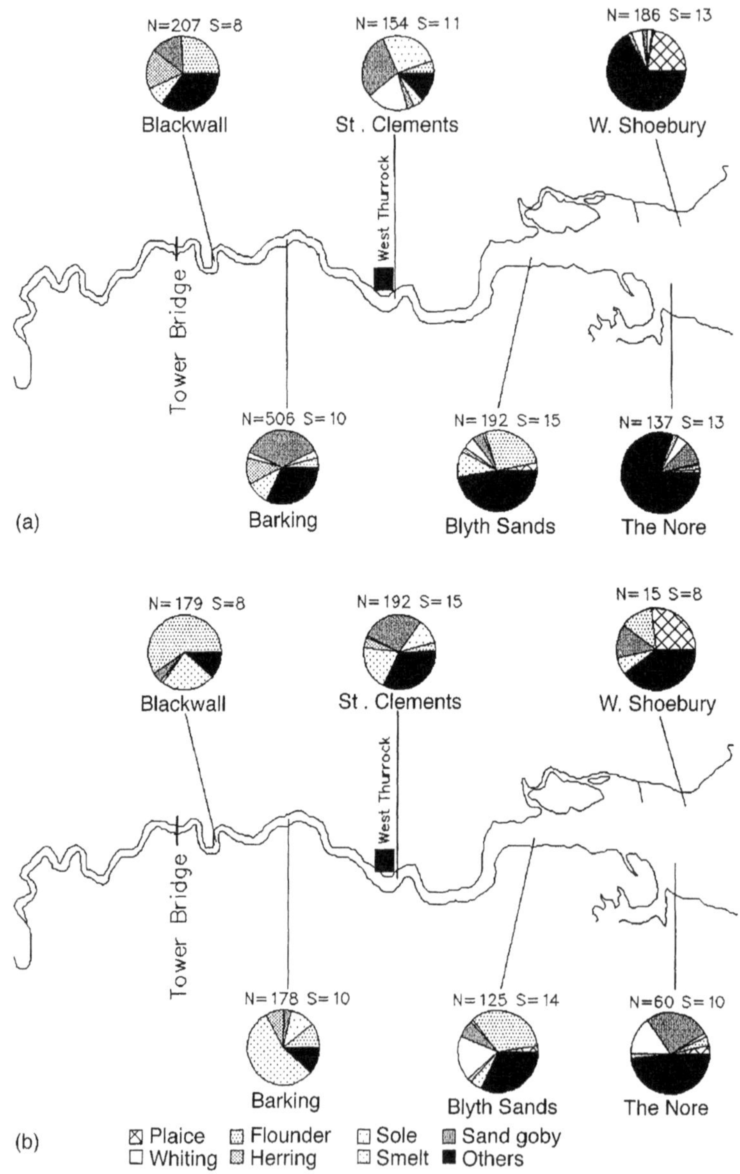

Figure 7.15 Trawling results for **(a)** 4th quarter, 1991, and **(b)** 1st quarter, 1992. The category 'Others' refers mainly to sprat, dab and pogge at the outer estuary sites, and sprat and eels at the inner estuary sites.

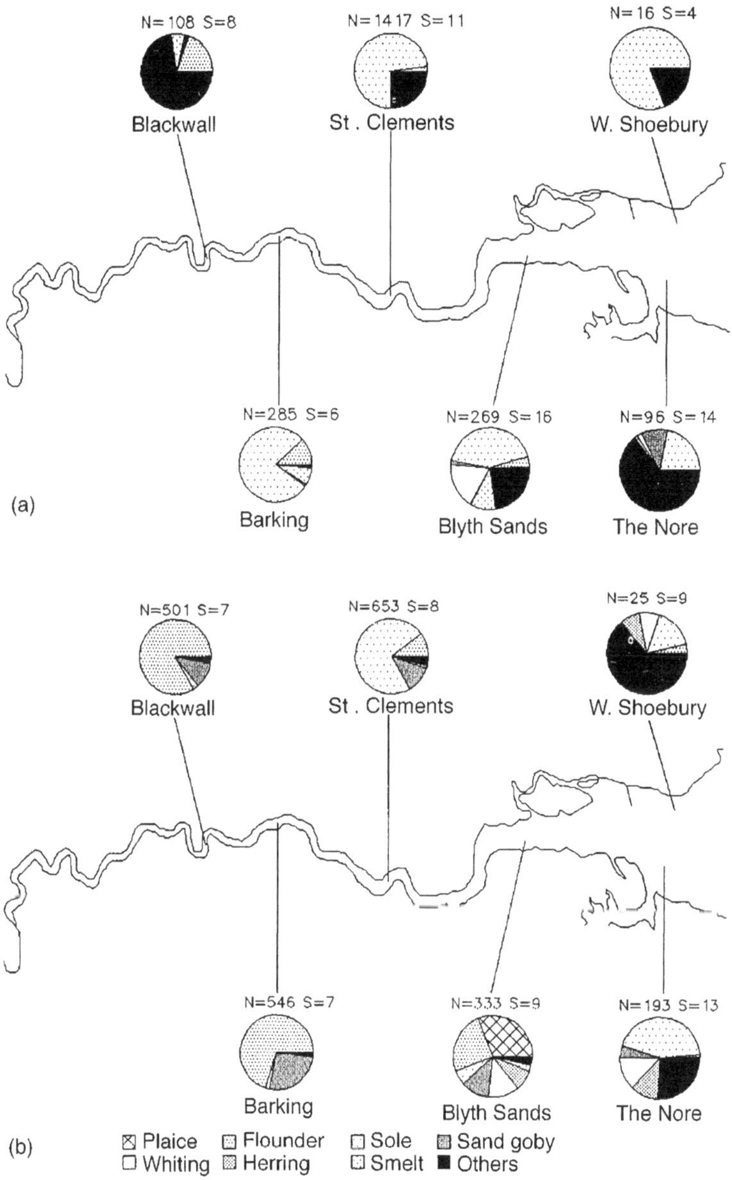

Figure 7.16 Trawling results for **(a)** 2nd and **(b)** 3rd quarters, 1992. The category 'Others' refers mainly to sprat, dab and pogge at the outer estuary sites, and sprat and eels at the inner estuary sites.

The seasonal migration of marine species was now largely complete, hence herring and whiting were only present in very low numbers at one or two sites. Numbers of sole increased dramatically as juveniles entered the river. Only a single plaice was recorded (at Blyth Sands), suggesting an offshore migration of this species during this quarter. Flounder abundance was similar to that in the previous quarters, while smelt continued their spawning run to the upper river and were only present in decreased abundance at a few sites.

(d) Third quarter of 1992

Twenty species were recorded during this quarter, 13 of which were recorded at The Nore.

Marine species such as plaice, sole, whiting, herring and pogge began to gather in the outer estuary (West Shoebury, The Nore and Blyth Sands) prior to movement further into the estuary. Sole were still present in large numbers, but appeared to be beginning a seaward migration. Flounder were now in very high numbers, due to the influx of juveniles into the estuary. Smelt were virtually absent and were likely to be distributed towards the upper reaches of the estuary.

7.8 SUMMARY

Fish species of the Thames estuary can show considerable variation in annual and/or seasonal abundance and distribution, most of which is due to natural migrations to and from spawning, feeding and nursery areas.

After a period of recovery, the fish community has now reached a state of fragile equilibrium which can be perturbed by extremes in environmental parameters. Consideration of the fish community as a whole, therefore, is likely to be of continued use for monitoring the effects of water quality on the biology of the Thames. Further work is required to establish a clear understanding of the effects of human activities upon the fish communities in the Thames estuary. It is important that links between water quality measurements and fish community characteristics are examined to determine the current relative condition of the fish community/populations. The EA and others have collected extensive data on a number of water quality parameters and, as this chapter has shown, there is a great deal of information available concerning the status of the fish populations. Individual species, such as flounder and smelt, have also been identified as possible biological tools for water quality monitoring. Further study, particularly of the former, is required to assess their suitability in this role.

A number of potentially important factors may be having highly significant impacts on the estuarine ecosystem and need further study.

Perhaps the most important of these is the fate and role of persistent contaminants (i.e. heavy metals, pesticides) and their potential ecological effects upon species populations and bioaccumulation within estuarine food chains.

The assessment of these additional factors, when combined with the existing information, will allow a better understanding of the complexities of the Thames estuary environment and enable efficient management plans to be developed to mitigate current and future problems.

CHAPTER 8

Host–parasite interactions: case studies of parasitic infections in migratory fish

Margaret Munro, Phillip Whitfield and Steve Lee

8.1 INTRODUCTION

The previous chapters have shown how the improved disposal of human and industrial wastes over the last two decades has led to a gradual improvement in the water quality of the Thames estuary, and the subsequent recolonization by fish, invertebrates and algae.

The re-establishment of the estuarine ecosystem, with its complex hydrology and salinity-delineated food webs, has resulted in the recreation of a vast range of ecological niches suitable for parasitic organisms. Many of these niches have since been colonized and the estuary now supports an abundant and diverse parasitic fauna, including microparasites such as viruses, bacteria, fungi and protozoans, and macroparasites that include crustaceans (mainly copepods) and helminths. The helminth (or worm) taxa identified to date include three groups of platyhelminthes: the Monogenea (ectoparasitic or external flukes), the Digenea (endoparasitic or internal flukes) and the Cestoda (tapeworms); and two other phyla: the Nematoda (roundworms) and the Acanthocephala (thorny-headed worms).

This diverse range of parasites utilizes a number of life cycle strategies. The simplest are the direct life cycles, which involve a single host species, and parasites employing this strategy were probably the first to recolonize the estuary when host species diversity was still relatively low. This type of life cycle occurs in most of the microparasite species, and in many nematodes and copepods. An example is the fish nematode *Cucullanus minutus*. Eggs of this nematode are voided in fish faeces and hatch on the

A Rehabilitated Estuarine Ecosystem. Edited by Martin J. Attrill.
Published in 1998 by Kluwer Academic Publishers, London. ISBN 0 412 49680 1.

estuary bed, releasing free-living, actively swimming larvae. These are thought to be directly infective to other definitive host fish. Newly acquired *C. minutus* embed themselves into the fish's gut wall, where they moult twice before entering the lumen of the intestine. In the intestine, they attach themselves to the gut wall, become sexually mature and reproduce (Mackenzie and Gibson, 1970).

A more complex indirect life cycle occurs in other nematode species and amongst the digenean, cestode and acanthocephalan parasites. These use a vertebrate definitive host, in which they undergo sexual reproduction, and one or more intermediate hosts that harbour parasitic larval stages. Both invertebrates and vertebrates may act as intermediate hosts; they provide a protected environment for larval growth and development, encystment and/or asexual reproduction. Larval transmission to the definitive host often relies on natural predator–prey relationships which exist within the estuarine food-chain.

Because these life cycles depend on a stable community of host species for their success, it seems likely that the indirect life cycle parasites were relative late-comers in the Thames estuary recolonization succession. For example, the digenean fluke *Cryptocotyle lingua* needs piscivorous birds, fish and littorinid molluscs in order to complete its life cycle. *C. lingua* larvae reproduce asexually in the common periwinkle *Littorina littorea*, producing free-living larval stages which then infect a range of second-intermediate host fish, including gobies and smelt. The encysted parasite is finally transmitted to larid birds such as the herring gull, *Larus argentatus*, when they ingest live fish or scavenge the estuarine shores for carrion.

Most parasites are highly microhabitat specific; that is, they congregate on a specific part of the host's body surface, or migrate to a particular organ. The ectoparasitic copepod *Lepeophtheiris pectoralis*, for example, commonly infects flounder in the lower estuary. This parasite clusters beneath the pelvic and upper pectoral fins of its host, whereas other species of copepod occur either attached to the gill filaments or inside the branchial cavity. Microhabitat specificity therefore not only facilitates mate location, but also extends the number and diversity of niches available for other parasite species. Niche separation may also occur by different parasites being prevalent at different times of the year. The two fish nematodes *Cucullanus minutus* and *C. heterochrous* have similar direct life cycles, and both live in the intestines of flounder. These nematodes enter the gut lumen from the surrounding tissues during the summer, usually between March and September. *C. minutus* then matures rapidly, reproduces, and dies off by the autumn, leaving the intestine to *C. heterochrous*, which then matures and remains in the lumen until the following year (Mackenzie and Gibson, 1970).

During the rehabilitation of the Thames estuary, the expanding parasitic fauna unfortunately attracted little attention; since the rehabilitation, studies of host–parasite systems have been few in number (Munro *et al.*, 1989; Lee and Whitfield, 1992; Munro, 1992; Rassai, 1992) and do not provide an even coverage of the various host taxa. These two gaps in our knowledge make it impossible to compile anything approaching a comprehensive list of the parasite species presently inhabiting the estuary, or to reconstruct the way in which parasite recolonization occurred.

We have therefore attempted to illustrate the diversity of estuarine host–parasite relationships by describing the case histories of two highly prevalent parasitic infections which occur in the common estuarine fish, flounder (*Platichthys flesus*) and smelt (*Osmerus eperlanus*). The ecological complexity of an indirect life cycle macroparasite infection is illustrated by the results of our recent epidemiological surveys on the fish acanthocephalan *Pomphorhynchus laevis*. The second part of the chapter then describes the ultrastructure and epidemiology of a simple direct life cycle microparasitic infection: virus-associated spawning papillomatosis, which occurs in migratory smelt.

8.2 THE INTESTINAL FISH HELMINTH *POMPHORHYNCHUS LAEVIS*

8.2.1 Introduction

We first discovered *Pomphorhynchus laevis* in the Thames estuary in 1986, when flounder from the upper estuary were found to contain large numbers of intestinal worms. The parasite was later identified as belonging to the estuarine strain of *P. laevis* (Munro *et al.*, 1989), the least known of three strains of *P. laevis* to be found within the British Isles. The strains are known as English freshwater, Irish freshwater and estuarine *P. laevis*, and are distinguished purely on ecological grounds, such as non-overlapping geographical distributions and differential host utilization (Kennedy *et al.*, 1978, 1989; Kennedy, 1984). Our recent comparative studies on the English freshwater and estuarine strains have revealed intense similarities in both stable morphological characteristics (of which there are few), and in genetic constitution (Munro *et al.*, 1990; Munro, 1992). In short, despite well defined ecological differences, there is at present insufficient evidence of divergence to warrant the reclassification of the three strains as different species.

P. laevis is a dioecious and gutless endoparasite with an indirect two-host life cycle (Figure 8.1). Like other fish acanthocephalans, the parasite has a low definitive host specificity, and adults can live in the intestines of a wide range of freshwater, euryhaline and marine fish species. The worms bore through the intestinal wall of the host using an anterior attachment complex consisting of a spined proboscis (Figure 8.2b), with a basal swelling known as the proboscis bulb. When the proboscis has fully

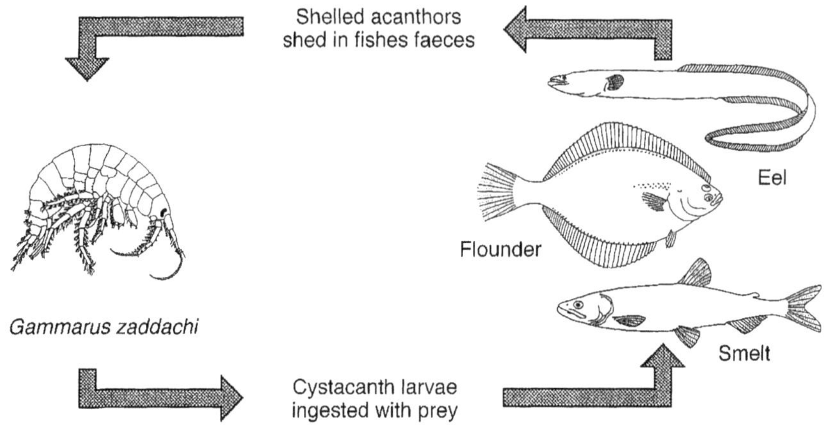

Figure 8.1 The life cycle of the estuarine strain of *Pomphorhynchus laevis* in the Thames estuary. Many other fish species, in addition to those illustrated, act as definitive host to the parasite. Fish redrawn from Wheeler (1978); *Gammarus zaddachi* redrawn from Lincoln (1979).

penetrated into the fish's peritoneal cavity, the bulb inflates and the worm is permanently attached to its host's intestine. In many fish hosts, *P. laevis* is also found in the peritoneal cavity and embedded in organs such as the liver and gonads. In these extra-intestinal locations, the parasite becomes encapsulated by host-derived fibrous tissue and eventually dies.

In contrast, there appears to be no lethal host immune response against intestinal *P. laevis*. The hind-bodies or trunks of intestinal worms measure about 10 mm long in flounder and lie freely in the lumen of the intestine (Figure 8.2a). The trunks contain the reproductive organs and, in mature inseminated females, large numbers of spindle-shaped eggs in all stages of development (Figure 8.2c). When fully embryonated, each egg, or shelled acanthor, is shed into the intestine and leaves the host in the faeces. Shelled acanthors are ingested by the scavenging amphipod *Gammarus zaddachi*, which acts as the main intermediate host to the estuarine strain of *P. laevis* in the Thames estuary. In the haemocoele of *G. zaddachi*, the acanthor larva develops into the final resting larval stage, known as the cystacanth, in about two months at summer water temperatures. Cystacanths are externally featureless larvae, closely resembling the adult parasites except for their fully inverted proboscis. They measure about 6 mm long, and have fully formed reproductive organs, with the testes of many males already containing threadlike spermatozoa. The large size and sexual precocity of these long-lived resting larval stages suggests that, once ingested by a suitable piscine host, establishment and reproduction would occur in rapid succession.

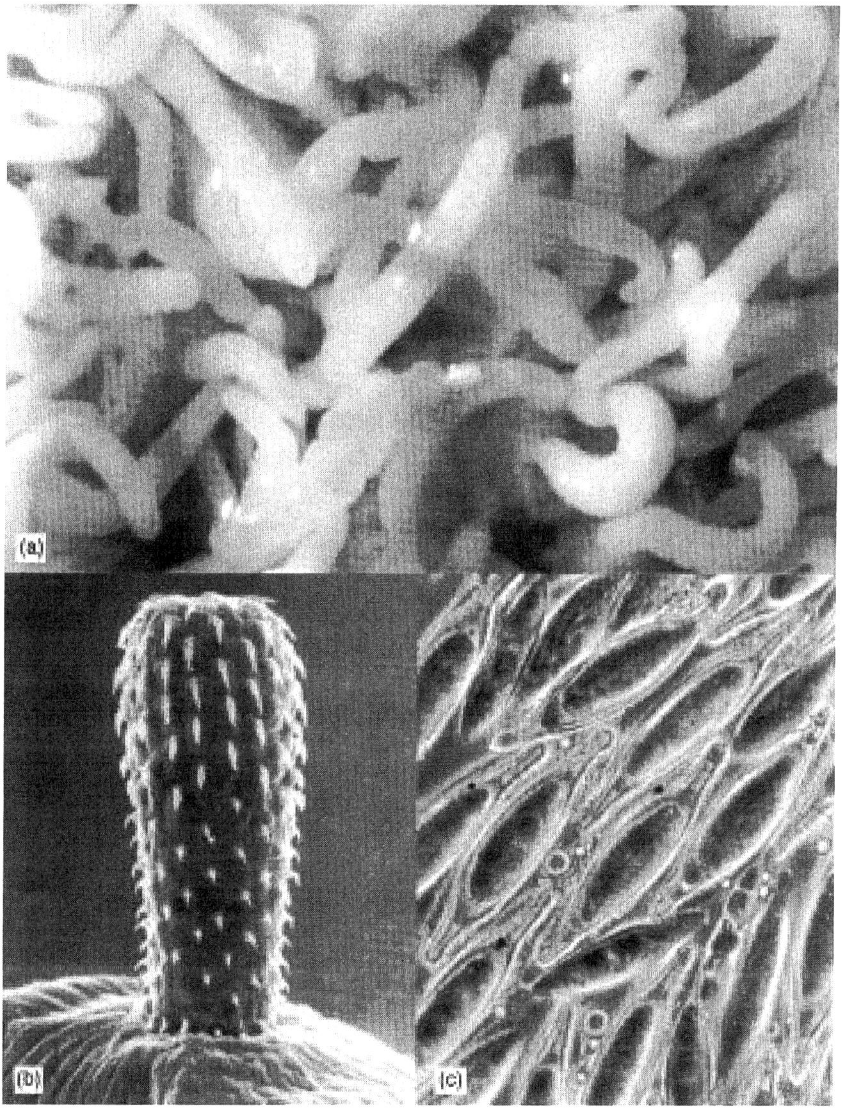

Figure 8.2 (**a**) Estuarine *Pomphorhynchus laevis* attached to the posterior intestine of an age 1+ flounder from the freshwater reaches of the Thames estuary at Fulham. (**b**) Scanning electron micrograph of the anterior spined proboscis. (**c**) Mixed-age egg sample taken from the pseudocoelomic cavity of a mature female.

8.2.2 Infection patterns in the intermediate host

In the estuary, the life cycle of *P. laevis* is completed mainly within the salinity-delineated distribution of *G. zaddachi*, and the pattern of infection within the fish populations is essentially controlled by the movement of potential hosts through this estuarine region.

In the Thames estuary, *G. zaddachi* has an exceptionally broad distribution and has completely replaced *G. pulex* in the upper freshwater reaches of the estuary over the last 15 years (Andrews, 1977; Attrill *et al.*, 1996a). *G. zaddachi* is now found from the limit of the tideway at Teddington, through central London, to Woolwich (Chapter 6). In the springtime, populations are dispersed by the increased freshwater flow and *G. zaddachi* is found as far downriver as West Thurrock in Essex, where it often cohabits with *G. salinus* until the early summer. This region of the estuary, stretching some 60 km, represents the main infection zone of *P. laevis* in the tideway.

Within this region, we found that the prevalence or proportion of *G. zaddachi* infected with *P. laevis* cystacanths changed with the local salinity regime. In samples taken in April 1989, only 2% of *G. zaddachi* from the freshwater reaches of the estuary, at Battersea, contained *P. laevis* cystacanths. The prevalence increased to 9% in the City of London, which lies on the borderline of seawater incursion, and then fell back to 7% at West Thurrock, where the daily incursion of seawater results in large diurnal salinity changes. At the seaward end of its distribution, *G. zaddachi* gives way to its more salinity tolerant relative *G. salinus*, and this is the dominant amphipod species downriver from West Thurrock (Chapter 6). Infected *G. salinus* were not found at West Thurrock in the sample taken in April 1988.

In addition to this longitudinal estuarine pattern in intermediate host parasitization, we also found a seasonal trend in *P. laevis* prevalence, which appears to be controlled predominantly by the water temperature cycle (Figure 8.3). A one-year survey of *G. zaddachi* from the freshwater estuary at Battersea showed that prevalence peaked in August 1989, when almost 15% of the amphipods contained viable *P. laevis* cystacanths. In this same month, a 2% prevalence was recorded in *G. salinus* from West Thurrock, indicating that this species can also act as intermediate host under optimum environmental conditions.

As infected *G. zaddachi* were found in all population samples, it seems likely that the transmission of cystacanths to predatory fish occurs throughout the year, although the transmission rate will be affected by both the seasonal prevalence of infection in *G. zaddachi* and by the fishes' food intake. In upriver habitats, *P. laevis* transmission will probably be maximized during the summer, when high temperatures speed up larval growth and induce greater feeding activity in the host populations, and may cease completely in the coldest months when many fish species fast. Further downriver, at West Thurrock, transmission will generally only occur during the Spring, when infected *G. zaddachi* are washed downriver by heavy freshwater flows, and in the late summer, when low numbers of infected *G. salinus* occur.

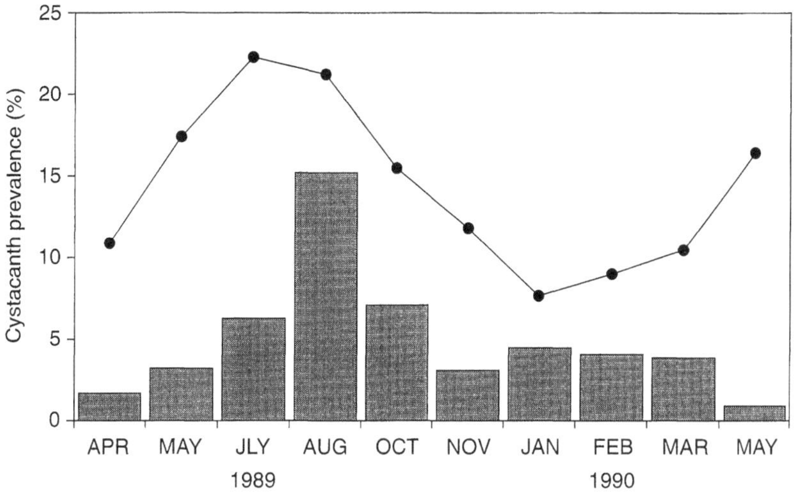

Figure 8.3 Seasonal patterns in *Pomphorhynchus laevis* cystacanth infection in six-weekly samples of *Gammarus zaddachi*, collected from the Thames at Battersea, London SW11, between April 1989 and May 1990. Bars = prevalence of *P. laevis* infection (%); solid line = mean monthly water temperatures (°C) recorded by the EA.

As well as these environmentally controlled spatial and seasonal differences in estuarine infection levels, we found that *P. laevis* cystacanths were not distributed randomly within the *G. zaddachi* population. Recently hatched *G. zaddachi*, less than 5 mm in length, were probably younger than the minimum cystacanth development time, and were not found infected with mature larvae (Figure 8.4). In larger (and therefore older) *G. zaddachi*, the prevalence of cystacanth infection increased with body length, as the populations gradually acquired infection through time. This association between host size and infection level means that fish preferentially predating different sizes of *Gammarus* will experience different rates of *P. laevis* transmission.

Figure 8.4 also shows that the cystacanths were distributed non-randomly between male and female *Gammarus*. Overall, 81% of the infected adult *G. zaddachi* were male, compared with an environmental sex ratio of 62% male. This anomaly appears to be related to host size: *G. zaddachi* shows well developed sexual dimorphism and females are significantly smaller than males. As *P. laevis* cystacanths are extremely large parasites compared with the size of their amphipod hosts, the smaller females may impose spatial or nutritional restrictions which either inhibit cystacanth development or which kill female amphipods, thus removing them from the population. In infected females, there was a positive correlation between their body length and the length of the cystacanth parasite, which did not exist in the infected male cohort. This

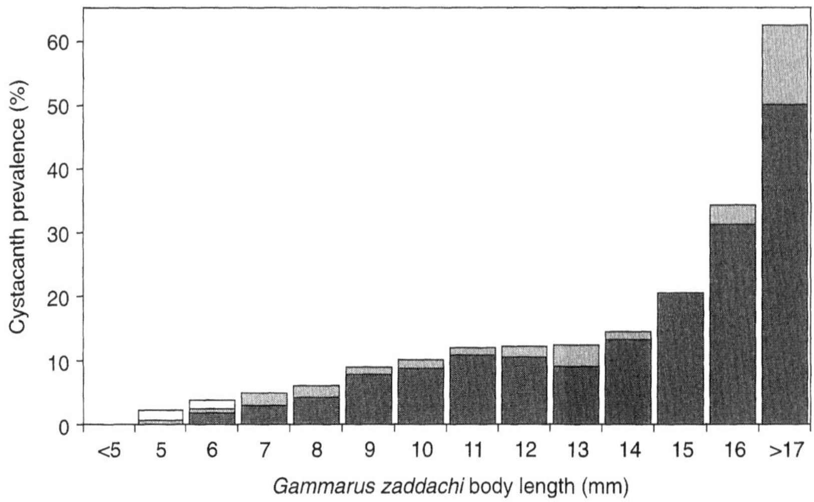

Figure 8.4 The prevalence of *Pomphorhynchus laevis* infection in relation to the body length of *Gammarus zaddachi*. Open bars = juvenile hosts; light shaded bars = female hosts; dark shaded bars = male hosts.

strongly suggests that female host size and growth rate are limiting factors in the development of these parasitic larvae.

8.2.3 The range of definitive host species

For a one-year period from April 1989, we also monitored the occurrence of adult estuarine *Pomphorhynchus laevis* in the Thames estuary fish populations from Lots Road Power Station in Fulham and from West Thurrock Power Station in Essex. We found *P. laevis* in 16 of the 32 fish species sampled (Table 8.1), suggesting that the parasite can establish and survive in any species of fish that passes through the estuarine infection zone and that feeds on *Gammarus*. This is by no means a comprehensive list of the fish hosts of *P. laevis* in the Thames estuary: about 60 fish species are recorded in the estuary every year (Chapter 7) and many of the less abundant unexamined species may also harbour *P. laevis*. Not all fish, however, provide the parasite with optimum physiological conditions for growth and reproduction, and gravid female *P. laevis* were only found in seven of the 16 infected fish species.

Similar patterns of host utilization are found in the English freshwater strain of *P. laevis*, which has been extensively studied in the River Avon, Hampshire (Hine and Kennedy, 1974; Kennedy *et al.*, 1978). In the Avon, infected fish species can be divided into three categories according to their suitability as definitive hosts for English freshwater *P. laevis*. In the preferred group of hosts, the parasites grow to full size and are reproductively active throughout the year. In the second host category, the parasites

Table 8.1 Known definitive hosts of *Pomphorhynchus laevis* in the Thames estuary, and reproductive status of intestinal parasite populations

Definitive host species	*Common name*	*Gravid female* P. laevis
Freshwater fish		
Abramis brama	Bream	+
Leuciscus leuciscus	Dace	++
Perca fluviatilis	Perch	+
Rutilus rutilus	Roach	–
Scardinius erythropthalmus	Rudd	–
Euryhaline fish		
Anguilla anguilla	Eel	++
Dicentrarchus labrax	Bass	–
Osmerus eperlanus	Smelt	+
Platichthys flesus	Flounder	++
Pomatoschistus microps	Common goby	–
Salmo salar	Salmon	–
Salmo trutta	Trout	–
Marine fish		
Agonus cataphractus	Hooknose	++
Ciliata mustela	Five-bearded rockling	–
Myxocephalus scorpius	Bull-rout	–
Solea solea	Sole	–

++, more than 20% of females gravid
+, between 1% and 20% of females gravid
–, gravid females not found

show some growth and reproduce at a suboptimal level, with small numbers of gravid females occurring, often only during the summer months. In the least favoured host group, the parasites can survive but fail to grow or reproduce.

The fish hosts of estuarine *P. laevis* can be similarly categorized on the basis of parasite size and reproductive activity. The preferred definitive host is undoubtedly the European flounder, *Platichthys flesus*. The flounder is the third most abundant fish species in the Thames estuary (Chapter 7) and between 1985 and 1991 accounted for 14.5% of all fish retrieved by the then NRA from the filter screens of West Thurrock Power Station. Over 95% of the flounder examined during our survey were infected with *P. laevis*, and parasite burdens ranged between one and nearly 500 worms; heavy infections such as that pictured in Figure 8.2a were commonplace. In flounder, mature gravid female *P. laevis* containing tens of thousands of eggs were abundant and were present all year round.

An example of a less favoured host species is the European smelt, *Osmerus eperlanus*. After flounder, this is the most common and the most heavily infected fish species found in the estuary. The abundance of smelt and flounder at West Thurrock Power Station between 1985 and 1991 was estimated by the Environment Agency at 21 449 smelt, and 34 534 flounder per 100 million gallons of cooling water (corrected figures, M. Thomas, personal communication), indicating the importance of these fish as hosts for *P. laevis*. Almost three-quarters of all the smelt examined were infected with *P. laevis*. In smelt the parasites were only about 6 mm long, the size of a cystacanth, suggesting that suboptimal intestinal conditions prevented *P. laevis* from growing to its full size. There were also very few gravid females and these produced relatively small numbers of eggs.

In other host species such as the sea bass, *Dicentrarchus labrax*, intestinal parasites were similar in size to those found in smelt but showed no evidence of reproductive activity. Juvenile bass enter the Thames estuary in the late autumn to overwinter, and many penetrate into the middle and upper reaches of the estuary, where they become infected with *P. laevis*. In recent years the numbers of bass in the estuary has been steadily increasing, probably due to the mild winters, and it is now the most abundant species after flounder and smelt (Chapter 7).

8.2.4 Infection patterns in flounder

Flounder spawn in the North Sea, generally between February and May (Wheeler, 1969b, 1978), and large numbers of fry migrate from the sea to the freshwater reaches of the Thames estuary during the summer months. In the initial stages of their migration, the young flounder are protected from *P. laevis* infection by their feeding habits. As we have seen, infective cystacanths are carried predominantly by large male *G. zaddachi*, and initially the fry take only small, newly hatched *Gammarus* in addition to foods such as copepods and diatoms. As they continue to migrate upriver, the young fish prey increasingly on larger food items, and hence start to acquire *P. laevis* from the larger infected *G. zaddachi*. This initial infection occurs during the first year of life, and the young flounder which reach the freshwater nursery ground around Fulham are all infected by the time they reach 1+ in age (Figure 8.5). In this region of the estuary, *G. zaddachi* is the major component of the macroinvertebrate fauna, and formed approximately 80% of the macrofauna biomass (g wet weight) at Cadogan Pier, Chelsea, in the year commencing April 1989 (Attrill, 1992). *G. zaddachi* hence forms the staple food of many fish species, including young flounder and smelt, resulting in the rapid accumulation of intestinal parasites.

There was no evidence of any seasonal patterns in the parasitization of these young flounder. Between April 1989 and May 1990, prevalence remained close to 100%, with only six of the 217 flounder examined being free from infection. No flounder or any other fish species were caught at Fulham between June and late September, in both 1989 and 1990. This was due to an annual, non-migrational movement of fish caused by poor local water conditions. There was also no seasonal trend in *P. laevis* intensity (the

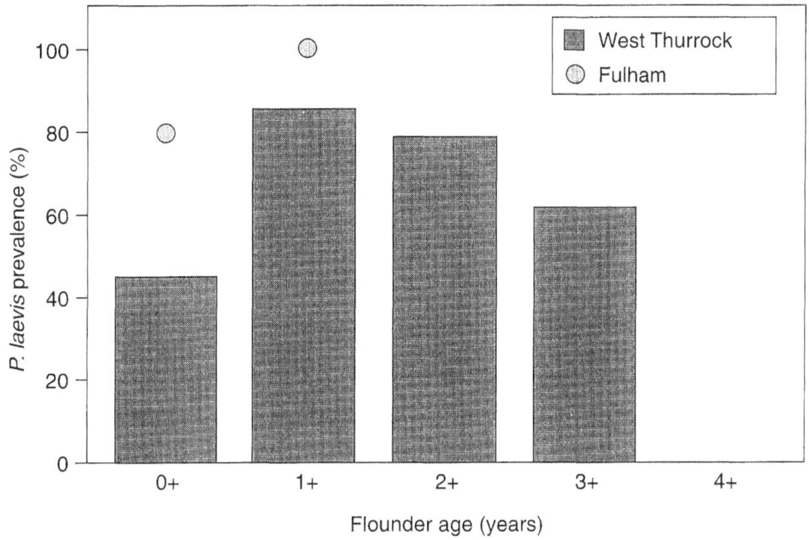

Figure 8.5 The host age-related prevalence of *Pomphorhynchus laevis* infection in flounder from the upper estuarine nursery ground in Fulham, London SW10, and from West Thurrock in Essex.

mean number of parasites per infected fish); intensity fluctuated irregularly throughout the year, with sample means ranging from 16 in November 1989, to 57 in February 1990, giving an overall average of 35 *P. laevis* per infected flounder.

When about 18 months to two years old, the flounder embark on their first spawning migration back to sea. During the initial stages of this migration, the flounder continue to acquire *P. laevis* as they pass through the estuarine infection zone. When the fish reach the middle and lower reaches of the estuary, the rate of parasite acquisition starts to decrease due to a combination of factors. Firstly, *G. zaddachi* eventually gives way to *G. salinus*, which carries only a very low level of cystacanth infection; secondly, the diversity of the macroinvertebrate fauna increases seawards (Attrill *et al.*, 1996b) and many fish cease to feed predominantly on *Gammarus*.

Not surprisingly, the West Thurrock flounder population was much more heterogeneous than that of the nursery ground, and consisted of both immature and mature fish ranging in age from 0+ to 4+. The flounder from West Thurrock were also far less wormy than their freshwater counterparts: 80% of the flounder examined between April 1989 and May 1990 were infected with *P. laevis* (compared with 97% at Fulham), and the average parasite burden was reduced, from 35 to 17 worms per infected fish.

Table 8.2 Level of *Pomphorhynchus laevis* infection in West Thurrock flounder infected with *Lernaeocera branchialis, Acanthochondria cornuta* and/or *Lepeophtheirus pectoralis*, compared with *P. laevis* infection in the cohort uninfected with these marine copepod ectoparasites

Copepod ectoparasites	P. laevis *prevalence (%)*	P. laevis *intensity*
Infected (n = 140)	69	10
Uninfected (n = 139)	91	22

At West Thurrock we found that infection was greatest in the age 1+ fish, as many of these would have acquired large parasite burdens in the upper estuary before returning to West Thurrock on their way to the sea. The prevalence of infection then decreased in progressively older flounder (Figure 8.5), due to age-related differences in prey choice and habitat. In older mature fish, which return to the lower estuary each summer after spawning, the decreasing prevalence levels indicated that the rate of parasite loss through senescence and death was greater than the rate of reinfection, suggesting that reinfection was not occurring at sea.

Two further pieces of evidence supported this theory. Firstly, about half of the West Thurrock flounder were also infected with a suite of ectoparasitic copepods, comprising *Lernaeocera branchialis, Acanthochondria cornuta* and *Lepeophtheirus pectoralis*. These marine crustaceans are intolerant of low water salinities and thus indicate that the infected flounder had recently migrated up to West Thurrock from more saline waters. It is significant, therefore, that the copepod-infected flounder showed lower levels of *P. laevis* infection than their predominantly non-migratory counterparts (Table 8.2).

In addition, we found that samples of coastal marine flounder taken from the North Sea off Suffolk, and from the Strait of Dover, Kent, showed low prevalence levels of between 6% and 15%, with most infected fish harbouring only two or three live intestinal worms. Many of the flounder that did not harbour live *P. laevis* contained the remains of dead parasites in the peritoneal cavity and viscera, indicating that they had once been infected. This suggests that the estuarine strain of *P. laevis* completes its life cycle exclusively in river estuaries, and not in the surrounding coastal waters.

8.2.5 A comparison of infections in flounder and smelt

Smelt are the commonest salmonid in the Thames estuary, and like the salmon, spawn in freshwater. During the winter, shoals of mature smelt migrate some 100 km upriver to their spawning grounds in the freshwater reaches of the Thames estuary.

We began an epidemiological survey in September 1990, when smelt, flounder and other resident fish species had returned to the Fulham area in response to the annual autumn improvement in water quality. Through the autumn, winter and early spring, the nursery-ground smelt population at Fulham consisted mainly of immature fish from the 1990 spawning. There were no seasonal patterns in the prevalence of infection in this community: prevalence remained above 70% between September 1990 and February 1991, and fluctuated irregularly between 96% (September, 1990) and 73% (October, 1990). In March, the first shoals of spawning adult fish arrived to swell the freshwater smelt population, and the prevalence dropped during the spawning season to 63% in March and then to 50% in April and May. By June, the entire fish population had again moved away from the Fulham area to avoid poor water conditions. At this time any remaining adult smelt would begin their migration back to sea, and the immature fish presumably moved downriver until the autumn.

Like the flounder population, the smelt showed size-related (and therefore age-related) infection patterns which resulted from their migratory movements through the *P. laevis* estuarine infection zone. The essentially non-migratory, immature fish generally measured less than 12 cm long, and showed high prevalence and intensity levels (Figure 8.6a,b). Like the young flounder, these fish feed predominantly on the locally abundant *G. zaddachi*, and hence become heavily infected with *P. laevis*. The largest of the immature smelt harboured an average of 29 *P. laevis* (Figure 8.6b), an infection level comparable to that observed in flounder from the same site.

The majority of the smelt measuring more than 12 cm long were spawning adults. These fish live in coastal waters, or around the estuary mouth, for most of the year and are probably only exposed to *P. laevis* infection during their annual spawning migration. Fish acanthocephalans seldom live for more than a year in their definitive hosts (Kennedy, 1985) and infections in adult smelt are therefore unlikely to persist from year to year. Because of these factors, the prevalence and intensity of *P. laevis* infection was lower in the adult smelt cohort than in the juveniles from the nursery ground (Figure 8.6a,b). The low levels of infection in the adults are also maintained by age-related changes in prey choice, with larger smelt preferring small fish to *Gammarus* and other invertebrates. Overall, infected smelt were far less wormy than flounder, and had an average parasite burden over the year of 15 *P. laevis*, compared with 35 in flounder.

Although they are the second most heavily infected species in the estuary, smelt are one of *P. laevis*'s less favourable definitive hosts and the parasite fails to attain its full reproductive potential in this fish. In flounder, we found that half of the female *P. laevis* were gravid and that the average number of eggs in the pseudocoelome was 29 000, with 35% being fully embryonated shelled acanthors (Table 8.3).

Smelt had comparatively few gravid females and these produced only small developing egg populations containing few shelled acanthors. During our one-year surveys, each infected flounder harboured on average a total of 247 000 *P. laevis* eggs, of which 86 000 were shelled acanthors, ready to be

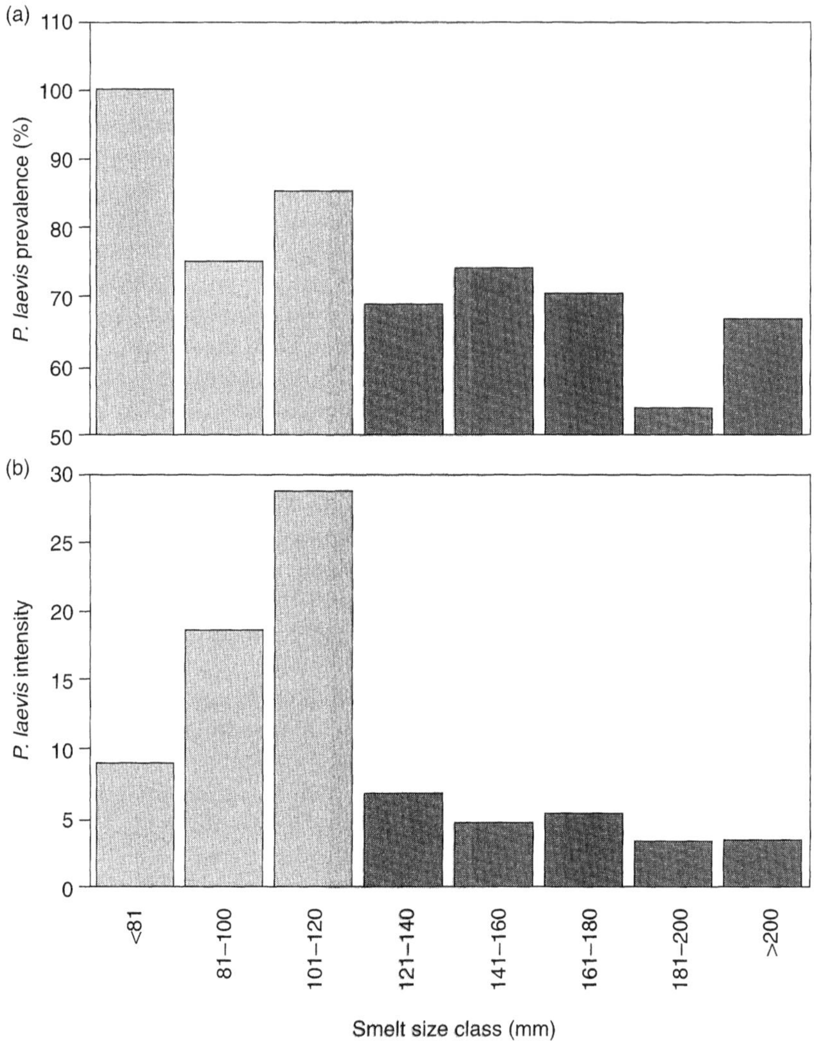

Figure 8.6 The relationships between *Pomphorhynchus laevis* infection level and smelt fork length (mm): **(a)** prevalence (%), and **(b)** parasite intensity. Bars with light shading represent samples consisting predominantly of immature smelt; those with dark shading represent samples consisting mainly of spawning adults.

shed into the environment. The average infected smelt contained only about 900 eggs, with only five of these being fully mature.

It seems likely that this massive reduction in the reproductive success of *P. laevis* in smelt results from a partially effective host immune response – one which allows the parasite to survive but stunts its growth and reproductive

Table 8.3 A comparison of the reproductive output of intestinal *Pomphorhynchus laevis* populations infecting flounder and smelt from the Thames estuary

Pomphorhynchus laevis	*Platichthys flesus*	*Osmerus eperlanus*
Prevalance (%)	97	74
Intensity	35	15
Females gravid (%)	50	10
Mean no. eggs/female	29 000	1200
Shelled acanthors (%)	35	0.5

rate. Apart from being smaller and less reproductively active, smelt *P. laevis* differ from the flounder parasites in their coloration. In flounder, all *P. laevis* are a homogeneous creamy-white, but in smelt the parasite often shows bright orange pigmentation in the areas in direct contact with the host's tissue – typically the neck which spans the gut wall, and occasionally the proboscis and bulb. This coloration, which is presumed to be caused by carotenoid deposits as these colour the English freshwater strain of *P. laevis* (Barrett and Butterworth, 1973), may represent another facet of a complex debilitating host immune reaction which is absent in the flounder.

8.2.6 Pathogenicity to estuarine fish

Adult acanthocephalan infections are rarely the cause of serious disease epidemics in wild animal communities, although many isolated fatalities have been attributed to the group (Nickol, 1985). In the wide range of fish species infected with *P. laevis* in the estuary, immature flounder from the nursery ground harboured the largest parasite burdens, and hence potentially supported the most pathogenic infections. In these heavily infected young fish, a high degree of parasite aggregation resulted in one-fifth of the fish harbouring almost half of the parasite population. The parasite burdens in this heavily infected cohort ranged between 40 and 498 worms per flounder, and these fish might be expected to experience the highest degree of morbidity related to the associated gut pathology, potential malnutrition, and other unknown parasite-induced effects.

Given the ubiquity of infection within the immature flounder population, it was difficult to assess the degree of pathogenicity associated with *P. laevis*, as there is no age-matched uninfected flounder cohort to act as a control group. In addition, many of the fish harboured other parasite species, which, although less abundant, may also have deleterious effects on the host. Under these circumstances, the best estimate of parasite pathogenicity available to us was condition factor analysis. The condition factor, K, can be used to detect changes in the fish's allometric growth pattern when related to age, body length, or to parameters such as parasite intensity. K is derived from the following equation, which has been adjusted to account for the biomass of *P. laevis*:

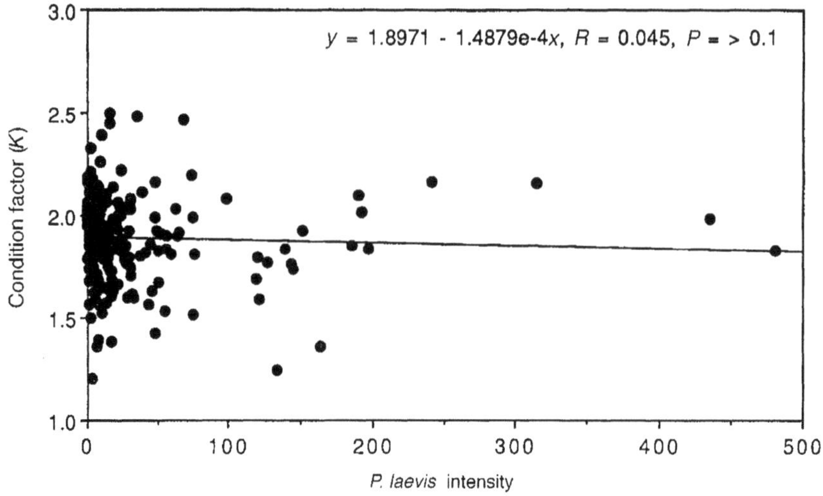

Figure 8.7 The relationship between the intensity of *Pomphorhynchus laevis* infection and the condition factor (K), in immature flounder from the Thames estuary at Fulham.

$$K = [(HW - PW)/L^3] \times 100,$$

where HW and PW are the wet body weights (g) of the host and of the total *P. laevis* burden, respectively and L is host body length (cm).

We reasoned that if there were any generalized systemic harm associated with *P. laevis* infections, which affected the relationship between fish length and weight, then its degree would be positively correlated with the size of the parasite burden. In fact, as Figure 8.7 shows, there was no significant trend of any sort between condition factor and *P. laevis* intensity.

This means that either the host's condition factor is similarly lowered by all levels of infection, ranging between one and nearly 500 worms, or, more reasonably, that *P. laevis* (and other parasite species) does not have an observable systemic impact on flounder condition. It is important to realize that this analysis only considers one somatic measure of condition, and it is conceivable that infections could have deleterious effects on other host attributes. They could, for instance, delay the onset of sexual maturation within the heavily infected immature flounder inhabiting the upper estuary. In smelt, where the level of *P. laevis* infection was low enough to provide infected and uninfected host groups, a condition factor analysis comparing smelt length to K value (Figure 8.8) failed to identify any significant parasite-linked differences in the allometric growth patterns of the two host cohorts (mean $K_{uninfected} = 0.83$; mean $K_{infected} = 0.86$).

In fact, the heavy parasite burdens observed in immature flounder and smelt do not persist throughout the fish's life span. Intestinal parasites are

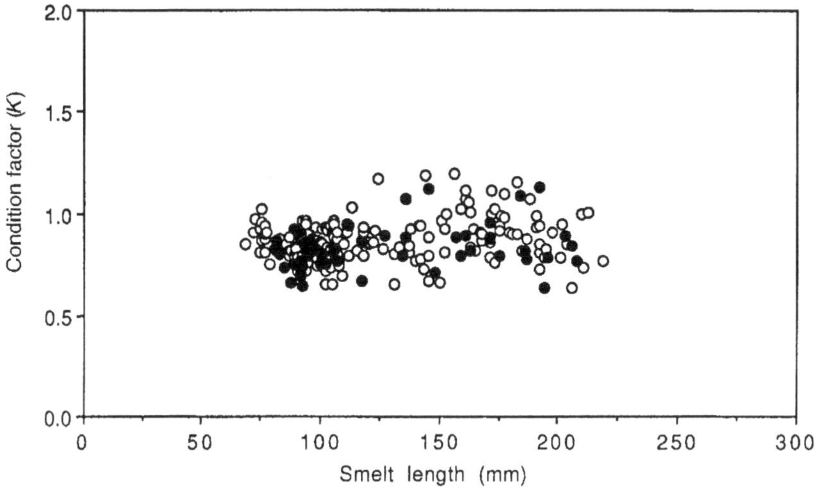

Figure 8.8 The relationship between the fork length (mm) of smelt from the Thames estuary and the condition factor (*K*): open circles represent smelt infected with *Pomphorhynchus laevis*; solid circles represent uninfected smelt.

gradually lost once the fish have migrated downriver to the lower estuary and surrounding coastal waters, and *P. laevis* is never again accumulated in such numbers. This temporally restricted infection pattern may alleviate any undetected deleterious effects, allowing any impaired growth patterns to normalize, and enabling natural healing processes to regenerate damaged areas of the gut wall.

8.3 VIRUS-ASSOCIATED SPAWNING PAPILLOMATOSIS IN SMELT

8.3.1 Introduction

Spawning papillomatosis is a disease of the European smelt, *Osmerus eperlanus*, which occurs in the Baltic Sea (Breslaeuer, 1916) and the River Elbe, West Germany (Anders and Möller, 1985; Anders, 1989). Prior to our survey, which first identified spawning papillomatosis in the Thames estuary in the Spring of 1989 (Lee and Whitfield, 1992), the only evidence of the disease in British waters was a single preserved smelt specimen dating back to the turn of the century, and forming part of the James Johnstone collection of fish parasites and diseases. This smelt, currently held at the Merseyside County Museum in Liverpool, bears white ovoid fin tumours which are macroscopically similar to those found infecting smelt in the Thames estuary today.

8.3.2 Morphology and ultrastructure

In the spring of 1989, about one-fifth of smelt examined from the upper reaches of the Thames estuary showed the smooth, white, hemispherical fin tumours characteristic of spawning papillomatosis. Similar tumours were occasionally found on the head and body of infected fish. In histological sections, the fin papillomas, which measured approximately 2.5 mm in length (Figure 8.9a), exhibited an internal structure comprising a non-vascular outer layer of tightly packed transformed epithelial cells, and an inner loosely packed necrotic region.

TEM sections revealed that virus particles characteristic of the herpes virus group were present in over 75% of fin tumour cells. All infected cells contained intranuclear herpes virus particles with icosahedral protein capsids measuring 95–100 nm in diameter. Some of these had darkly staining cores (Figure 8.9b), whilst others appeared to be hollow, suggesting that viral replication was occurring within the nucleus, and that some capsids already possessed their double-stranded linear DNA core.

In addition, herpes virus particles were commonly found in the cytoplasm of tumour cells, and also associated with the nuclear envelope. The cytoplasmic particles were larger (about 150 nm diameter) and more irregularly shaped than the intranuclear ones, due to the capsid acquiring an external lipid bilayer envelope, possibly during its passage through the nuclear envelope. Particles lying between the inner and outer nuclear membranes measured about 130 nm in diameter and lacked the outer envelope (Figure 8.9c).

8.3.3 Epidemiology

Although we assume that spawning papillomatosis in smelt results from a direct life cycle oncogenic herpes virus infection, we know very little about the distribution of the virus within the fish community, whether latent infections occur, and what triggers oncogenic tumour-forming activity.

We started to monitor the smelt population at Fulham, which lies close to the upper estuarine spawning and maturation grounds, in October 1989 (Lee and Whitfield, 1992). We found that the seasonal changes in population density, body length and sexual maturity, which were related to the annual spawning migration, were also closely allied to the level of spawning papillomatosis.

During the winter of 1989, relatively low numbers of immature smelt inhabited the upper estuarine nursery grounds, and all were free of fin papillomas (Figure 8.10 a,b). In March, the size of the smelt population apparently increased 20-fold, due to the massive influx of spawning adults, and papillomatosis prevalence was at its highest, with 21% of the fish exhibiting one or more fin tumours. By April, the early spawners were already returning to sea and the prevalence had dropped slightly to 19%. During these two months, the number of tumours observed on

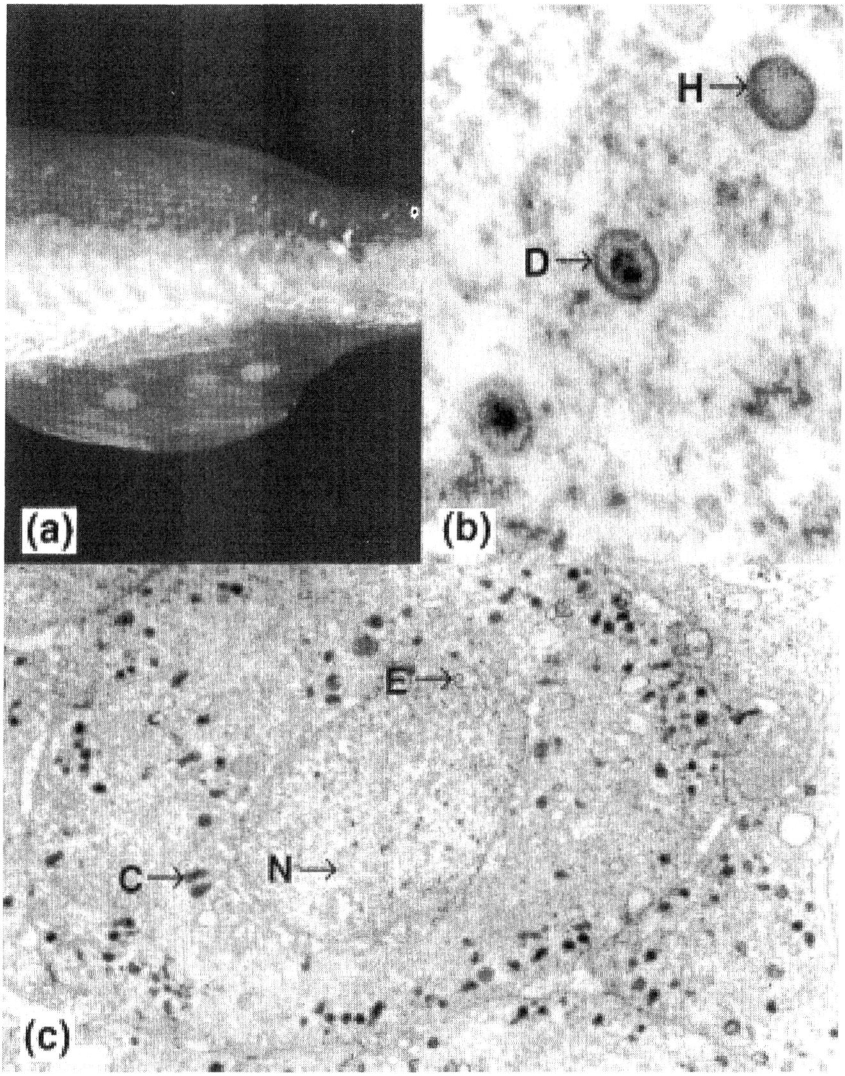

Figure 8.9 **(a)** The anal fin of a smelt showing three discrete ovoid tumours. **(b)** Icosahedral intranuclear herpesvirus particles: (H) hollow; (D) with darkly staining core. **(c)** A herpesvirus-infected tumour cell with intranuclear virus particles (N), nuclear envelope-associated particles (E), and irregularly shaped cytoplasmic particles (C).

papillomatous smelt ranged from one to 13, giving an average burden of four papillomas per fish.

Smelt samples taken in May showed that the majority of adult fish had already begun their return journey to the sea and that the disease was, at

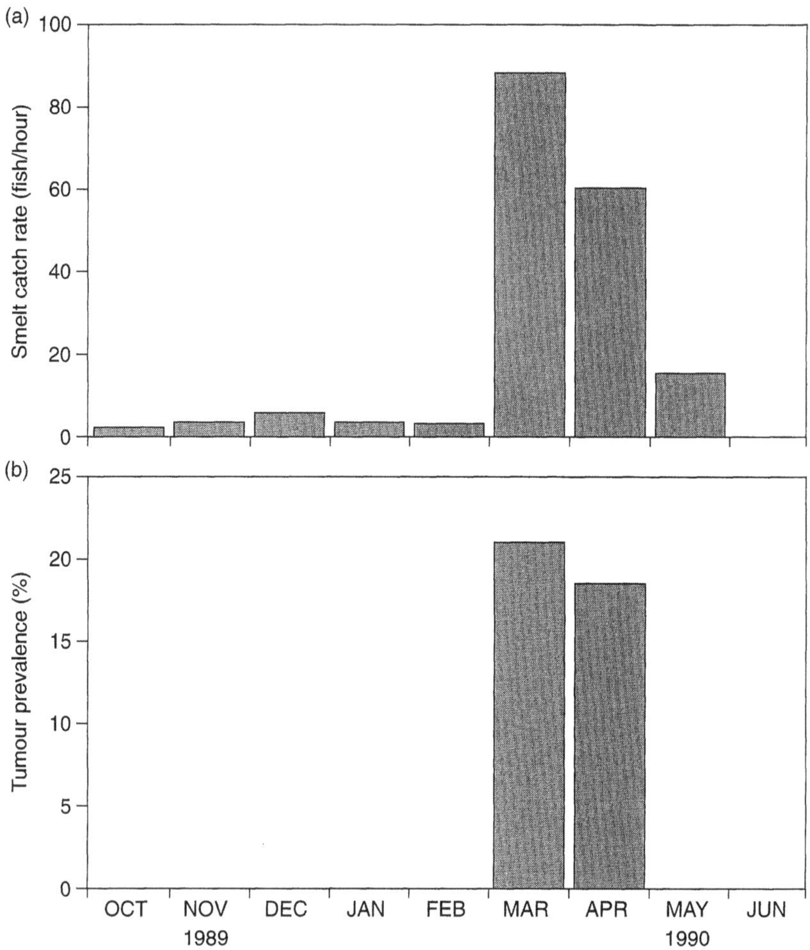

Figure 8.10 Seasonal patterns in **(a)** smelt abundance in the upper Thames estuary, demonstrated by catch rate (mean number of fish/hour), and **(b)** the prevalence of spawning papillomatosis.

least temporarily, eliminated from the upper estuary. The absence of smelt between June and September was due to the annual local deterioration in water quality which promotes the temporary movement of all resident fish species away from the sampling area. The displaced smelt population at this time would be expected to consist mainly of fry and immature fish, as it was in the autumn of 1990 when the fish returned.

In the Elbe estuary, a survey which followed the annual movements of mature smelt (Anders, 1989) revealed that the appearance of small nodular fin tumours coincided with the annual maturation of the gonads in the autumn. The tumours then developed during the winter and were most numerous at spawning time, when they also reached their maximum size. After spawning, the tumours became progressively easier to dislodge and had all fallen off by the end of May. Anders also found that the lesions left by tumour shedding were rapidly colonized by pathogenic bacteria, resulting in skin ulcerations. There was a close correlation between the seasonal prevalence patterns of spawning papillomatosis and skin ulceration, with the latter occurring after a one-month time-lag.

In the 222 Thames estuary smelt examined during March and April, tumour prevalence and intensity were both related to the size of the fish, with immature smelt less than 15 cm long being completely tumour-free (Figure 8.11). The smallest infected smelt, measuring less than 16 cm, probably represent the young adults in their first spawning season. In larger fish the prevalence of papillomatosis generally increased with host size (and presumably with age), and 45% of smelt over 20 cm long exhibited the disease. In the Elbe estuary, where there was a much greater spread in smelt size, fin tumour prevalence peaked at about 40% in fish 20 cm long and then declined steadily, with fewer than 10% of fish over 26 cm long being papillomatous (Anders, 1989).

In the Thames, larger papillomatous fish also generally had more tumours than smaller ones, possibly due to the accumulation of infected cell foci remaining from previous years' tumours. Such latent viral infections may predetermine the body surface distribution of tumours during successive spawnings.

The nature of possible environmental and physiological interactions which result in the manifestation of herpes virus-associated tumours are at present poorly understood (Möller, 1987). The simplest hypothesis is that latent viral infections are present in all smelt stocks, and that oncogenic activity is triggered by a combination of specific environmental and/or physiological stimuli. These may include spawning-related changes in host hormone levels, or physiological stress caused by the migration from sea to freshwater, or by exposure to different pollutant loads. There is at present no direct evidence for latent herpes virus infections within the smelt population, and if these do not exist, migratory fish may become infected (or reinfected) with the virus in the Thames estuary each spring. Patterns of parasitization similar to those described above have since been observed in the 1990 and 1991 Thames estuary smelt populations, which suggests that spawning papillomatosis may be a persistent feature of migratory smelt populations in the estuary.

Because spawning papillomatosis is one of several widespread parasitic infections identified in smelt, any deleterious effects resulting from this single disease are difficult to isolate and assess. Condition factor analysis of completely uninfected smelt, and of those infected with papillomas (and helminth parasites in many cases), yielded similar and overlapping data sets (mean $K_{uninfected} = 0.83$; mean $K_{infected} = 0.95$) (Figure 8.12). Spawning papillomatosis (and other parasitic diseases) therefore appears to have

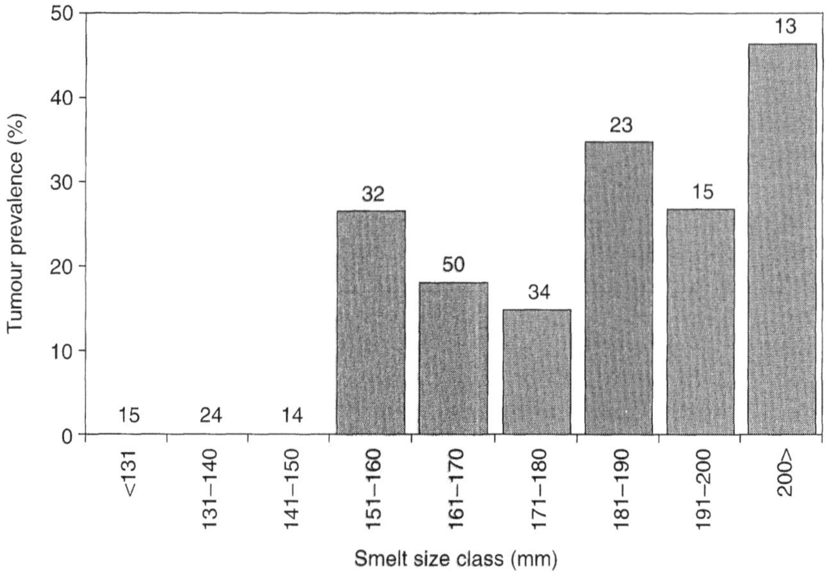

Figure 8.11 The relationship between smelt fork length (mm) and tumour prevalence (%). Numbers indicate the sample sizes in each smelt length class.

little effect on the allometric growth pattern of smelt in the Thames estuary. There was also no evidence of linkages between concurrent parasitic infections, and so the acquisition of helminth parasites such as *P. laevis* appeared not to affect the fishes' likelihood of developing spawning papillomatosis.

8.4 OTHER PARASITES OF MIGRATORY FISH

In addition to *Pomphorhynchus laevis* and herpes virus-associated spawning papillomatosis, several other parasite species were identified during the course of our epidemiological surveys. Of the two migratory fish examined, smelt had the narrowest range of parasitic fauna (Table 8.4).

This host harboured two digenean parasites (*Brachyphallus crenatus* and *Cryptocotyle lingua*), an adult direct life cycle nematode (*Hysterothylacium aduncum*) and encapsulated nematode larvae. The complex multi-host life cycle of *C. lingua* is described in the introduction to this chapter. In the Thames estuary, smelt act as a second intermediate host to *C. lingua*, and the encysted metacercariae are visible as small black pigment spots embedded in the surface epithelium of the body and fin rays.

Flounder harboured a wider range of parasites belonging to six parasitic groups, including the Cestoda (Table 8.5). Tetraphyllidean larvae were recovered from the intestines of a small number of flounder from both

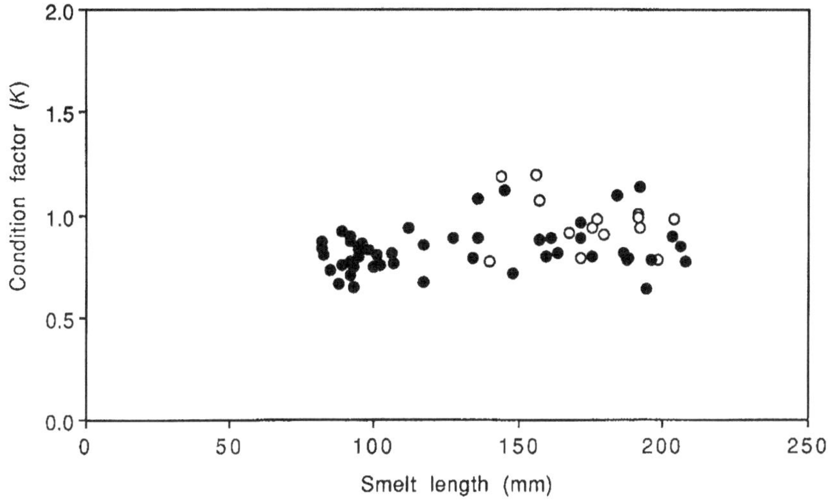

Figure 8.12 The relationship between the fork length (mm) of smelt from the Thames estuary and the condition factor (*K*): open circles represent smelt with virus-associated spawning papillomatosis; solid circles represent non-papillomatous smelt.

sampling sites. The Order Tetraphyllidea are small, highly specialized tapeworms which reproduce exclusively in elasmobranch fish. The life cycles are not well known, and the larvae have been identified in a range of intermediate host species including a variety of fish species, copepods and ctenophores.

In addition to unidentified encapsulated nematode larvae, three adult nematodes were commonly identified in the flounder gut: two species of *Cucullanus* (the life cycles of which are discussed in section 8.1) and *Paracapillaria*.

A second acanthocephalan species that was present in flounder was *Acanthocephalus anguillae*. Low numbers of *A. anguillae* were found

Table 8.4 Species of parasite infecting the European smelt (*Osmerus eperlanus*) in the Thames estuary

Parasitic group	*Species*	*Organs infested*
Viruses	Herpes-virus	Fin/body tumours
Digenea	*Brachyphallus crenatus*	Intestine
	Cryptocotyle lingua larvae	Body surface
Nematoda	*Hysterothylacium aduncum*	Intestine
	Unidentified larvae	Intestine/viscera
Acanthocephala	*Pomphorhynchus laevis*	Intestine/viscera

Table 8.5 Species of parasite infecting the European flounder (*Platichthys flesus*) in the Thames estuary

Parasitic group	*Species*	*Organs infested*
Microsporida	*Glugea stephani* spores	Intestinal wall
Digenea	*Cryptocotyle concava* larvae	Gill filaments
	Plagioporus varius	Intestine
	Zoogonoides viviparus	Rectum
Cestoda	Tetraphyllidean larvae	Intestine
Nematoda	*Cucullanus heterochrous*	Intestine/rectum
	Cucullanus minutus	Intestine
	Paracapillaria spp.	Intestine
	Unidentified larvae	Intestine/viscera
Acanthocephala	*Acanthocephalus anguillae*	Intestine/viscera
	Pomphorhynchus laevis	Intestine/viscera
Crustacea	*Acanthochondria cornuta*	Branchial cavity
	Lepeophtheirus pectoralis	Body surface
	Lernaeocera branchialis larvae	Gill filaments

cohabiting with *P. laevis* in the intestines and viscera of about 3% of the flounder examined from the upper estuarine nursery ground.

Three marine crustacean copepods (*Lernaeocera branchialis*, *Acanthochondria cornuta* and *Lepeophtheirus pectoralis*) were common in migratory flounder from West Thurrock and also in the coastal marine flounder samples taken from Suffolk and Kent. The most economically important of these, and the only one with an indirect life cycle, is *L. branchialis*. Flounder and other flatfish act as the intermediate host to this parasite. *L. branchialis* larvae settle and develop on the gill filaments, eventually producing adult male and female parasites which copulate before leaving the host. The males then die; free-swimming inseminated females locate a gadoid definitive host fish and metamorphose into a large, blood-sucking, worm-like form which produces a succession of filamentous egg strings (Whitfield *et al.*, 1988). These females are highly pathogenic: a single *L. branchialis* can reduce the body weight of whiting, *Merlangius merlangus*, by 5–10%, and heavier infestations have resulted in a weight losses of up to 42% (Kabata, 1970). As this weight loss is predominantly from the muscle, the parasite represents a serious threat to commercial gadoid fisheries in the North Sea.

A recent survey on the helminth parasitic fauna of eels (*Anguilla anguilla*) from the Thames estuary (Rassai, 1992) revealed that *P. laevis* was also the dominant parasite of this common catadromous fish in the tideway. In total, the 949 eels examined were infected with 17 species of helminth (Table 8.6), including *Brachyphallus crenatus*, *Hysterothylacium aduncum* and *Acanthocephalus anguillae* (mentioned earlier).

Table 8.6 Species of helminth parasite infecting the eel (*Anguilla anguilla*) in the Thames estuary

Parasitic group	*Species*	*Organs infested*
Monogenea	*Dactylogyrus* sp.	Gills
	Pseudodactylogyrus anguillae	Gills
	Pseudodactylogyrus bini	Gills
Digenea	*Brachyphallus crenatus*	Stomach
	Deropristis inflata	Intestine
	Helicometra fasciata	Intestine
	Podocotyle atomon	Intestine
	Podocotyle reflexa	Intestine
Cestoda	*Bothriocephalus claviceps*	Intestine
Nematoda	*Anguillicola crassus*	Swim bladder
	Goezia sp.	Stomach
	Hysterothylacium aduncum	Intestine
Acanthocephala	*Acanthocephalus anguillae*	Intestine
	Acanthocephalus clavula	Intestine
	Acanthocephalus lucii	Intestine
	Pomphorhynchus laevis	Intestine
	Echinorhynchus truttae	Intestine

8.5 SUMMARY AND CONCLUSIONS

This chapter has attempted to describe both the diversity and the ecological complexity of the parasitic fauna infecting the fish of the Thames estuary. In doing so, we have listed the parasites of only two of the 112 fish species identified in the estuary since 1964 (Chapter 7). Of the 18 parasites species infecting these hosts, we have sufficient knowledge to describe only *Pomphorhynchus laevis* infections and virus-associated spawning papillomatosis in some detail.

A much more satisfactory measure of parasite species diversity, and of long-term changes in infection patterns, could obviously have been obtained if the parasitic fauna had been monitored during the estuary's rehabilitation, when food webs were being rebuilt and when the returning host fish and bird species were constantly increasing the number of available ecological niches. Not only has this valuable opportunity been missed, but also, since the rehabilitation, few studies on the estuarine parasitic fauna have been carried out and the vast majority of fish and bird species in the estuary remain unexamined in this context.

The estuarine strain of *Pomphorhynchus laevis* is by far the most prevalent fish macroparasite in the estuary, and accounted for the bulk of the helminth biomass in all the fish species examined. Approximately 14 000 worms were recovered from the 750 flounder and smelt examined during our surveys. *P. laevis* illustrates the dependence of indirect life cycle parasites on a stable ecosystem containing self-sustaining intermediate and definitive host populations. The parasite is enormously successful in the Thames estuary and probably infects all of the marine, euryhaline and freshwater fish species that penetrate the middle and upper estuary and that feed predominantly on *Gammarus.*

The region in which *P. laevis* completes its life cycle is naturally controlled by the distribution of potential hosts within the estuary. During the 1970s two main factors were responsible for the upriver spread of the infection into near-freshwaters. Firstly, flounder gradually returned to their maturation grounds in the upper estuary, and in the spring of 1972 the first major upriver migration of flounder fry since the 1920s was recorded (Wheeler, 1979). An annual flounder migratory pattern in the Thames estuary was thus re-established in the 1970s, and this resulted in the influx of both infected and uninfected fish into the freshwater estuary. At this point in time, the resident amphipod species in the upper estuary was *Gammarus pulex* (Andrews, 1977). *G. pulex* is the intermediate host of the English freshwater strain of *P. laevis* (which is present in the non-tidal Thames and its freshwater catchment), but we do not know if *G. pulex* can act as intermediate host to estuarine *P. laevis*. If it can, then there is no reason why the parasite should not have become established in the upper tideway by the mid-1970s. If *P. laevis* depends on the presence of *G. zaddachi*, however, the region in which the parasite's life cycle could be completed would be limited by the distribution of this amphipod. *G. zaddachi* gradually displaced *G. pulex* in the upper estuary, probably during the 1980s, and eventually colonized the flounder and smelt nursery grounds near Fulham, and beyond this region to the limit of the tideway at Teddington. The annual influx of young flounder into the freshwater estuary provides the parasite with ideal conditions for the completion of its life cycle. These immature fish feed predominantly on *Gammarus*, are highly susceptible to infection, and support reproductively active adult worm populations.

Our example of an oncogenic microparasitic infection, herpes virus-associated spawning papillomatosis, shows that even relatively simple direct life cycles can be complicated by seasonal changes in the physiology and behavioural patterns of the hosts. The epidemiology of spawning papillomatosis will also potentially be influenced by the hosts' exposure to a range of estuarine conditions, and prey items, during the annual spawning migration from sea to freshwater.

Our analyses of the presence of macro- and microparasites in fish of the Thames estuary enables us to begin to differentiate between parasite infections and parasitic diseases in these estuarine hosts. Our data on prevalence and intensities are effectively describing infections: they are quantitative measures of the mere presence of parasites within fish populations. In addition, though, our attempts to measure systemic harm associated with infection by means of conditions factor analysis, confronts the issue of parasitic disease. In the absence of data on fish survival and reproductive success with and without parasites, it seems reasonable to suggest that an infection causes a disease state if systemic harm (rather than organ-localized pathology) is associated with its presence. Using condition factor data, we were not able to demonstrate the presence of parasitic disease in the estuary associated with either *P. laevis* infections in flounder or smelt, or with spawning papillomatosis in the latter species.

Finally, this chapter serves to highlight the amount of research needed to provide us with a working model of the complex parasite–host and parasite–parasite interactions that currently exist within the estuary. The gaps in our parasitological knowledge of the Thames estuary are most acute in the region of invertebrate hosts, and in the extensive bird communities that visit the shores of the Thames to feed.

CHAPTER 9

The Thames estuary saltmarsh plant and seagrass communities

Stephen Waite

9.1 INTRODUCTION

The total area of coastal saltmarsh in Britain is some 44 370 ha (Burd, 1989). Of the defined natural and semi-natural habitats remaining in Britain, only chalk grassland occupies a smaller area. The largest expanse of this saltmarsh occurs within the greater Thames estuary along the Essex coast. Some 4400 ha (10% of the total extent of British saltmarsh) is within the greater Thames, which may be taken as extending from Colne Point in Essex to Whitstable Bay in Kent. The largest concentrations are associated with Blackwater Bay (1102.85 ha) and the Medway estuary (754.46 ha) (Burd, 1989).

The gently sloping coast and high load of river silts entering the relatively sheltered waters of the greater Thames estuary favour the development of extensive areas of saltmarsh. These marshes form one of the largest areas of saltmarsh and intertidal mudflats in north-west Europe (Harmsworth and Long, 1986; Burd, 1989). They support substantial populations of wildfowl and wading-birds which graze the *Zostera* (seagrass or eelgrass) beds and feed on the large numbers of macro-invertebrates inhabiting the intertidal mudflats fringing the saltmarshes (Boorman and Ranwell, 1977; Harmsworth and Long, 1986) (Chapter 6). The productivity of these mudflats is undoubtedly linked to that of the neighbouring saltmarsh vegetation, which in Essex is estimated to produce 70 000 tonnes of organic matter a year (Harmsworth and Long, 1986). Since these east coast marshes, unlike those of the west coast, are subjected to little grazing, much of this production is exported by tidal action as litter

A Rehabilitated Estuarine Ecosystem. Edited by Martin J. Attrill.
Published in 1998 by Kluwer Academic Publishers, London. ISBN 0 412 49680 1.

and detritus to the surrounding mudflats. In addition to affecting macro-invertebrate productivity (Chapter 6), such inputs will have a major impact on the composition and productivity of meiofaunal communities.

9.2 SALTMARSH DEVELOPMENT

The ecology, formation and development of saltmarshes are authoritatively reviewed by Adam (1990) and Allen and Pye (1992). In addition, a very extensive bibliography of works on saltmarshes has been produced by Charman *et al.* (1986). The south-east coastline of Britain is complex and indented, formed essentially from of a series of drowned shallow river valleys. The major geological formation is the London Clay which is locally overlain to varying degrees of thickness with alluvium, Pleistocene deposits or gravel, producing a soft low-lying coastline where mudflats and saltmarshes are a dominant feature (Burd, 1992).

Saltmarshes may be defined as belts of higher plant vegetation which are periodically flooded by sea water, normally occurring between the mean high water neap tide level and mean high water. They require a supply of sediment and coastal conditions that promote accretion and protect building mudflats from erosion. Physico-chemical factors and the action of unicellular algae and bacteria cause fine particles to aggregate, increasing accretion and aiding the initial formation and stabilization of the mudflats. Until a mudflat is colonized by vegetation, it is easily eroded by storms. Once vegetated, the accretion rate increases; above-ground plant material reduces water velocity, causing sediment to accrete, whilst the roots bind and hold the sediments together. With time the marsh extends seaward, any point on the marsh surface increasing in elevation until either accretion is equal to erosion, or the point is no longer subject to tidal inundation. As the marsh builds, conditions will change as the extent of tidal inundation decreases. In response to these changes, a sequence of species becomes established and these are then displaced by later successional species. These changes are often reflected in the zonation of marsh vegetation (Ranwell, 1972; Randerson, 1979; Adam, 1990). Saltmarsh vegetation is frequently describe with reference to lower, middle and upper marsh zones (Figure 9.1), each zone tending to support a particular assemblage of plant species. The low marsh, or pioneer zone, starts around the mean high water neap tidal level; the middle marsh at about mean high water, and the upper marsh at mean high water spring tides (Long and Mason, 1983).

The expanse of saltmarsh depends on a fine balance between erosion and accretion. Changes in sediment load, tidal currents and occasional storm events may rapidly change the position of the seaward marsh limit. Saltmarshes frequently demonstrate cyclic episodes of expansion and contraction. However, the rate of east coast saltmarsh contraction is of

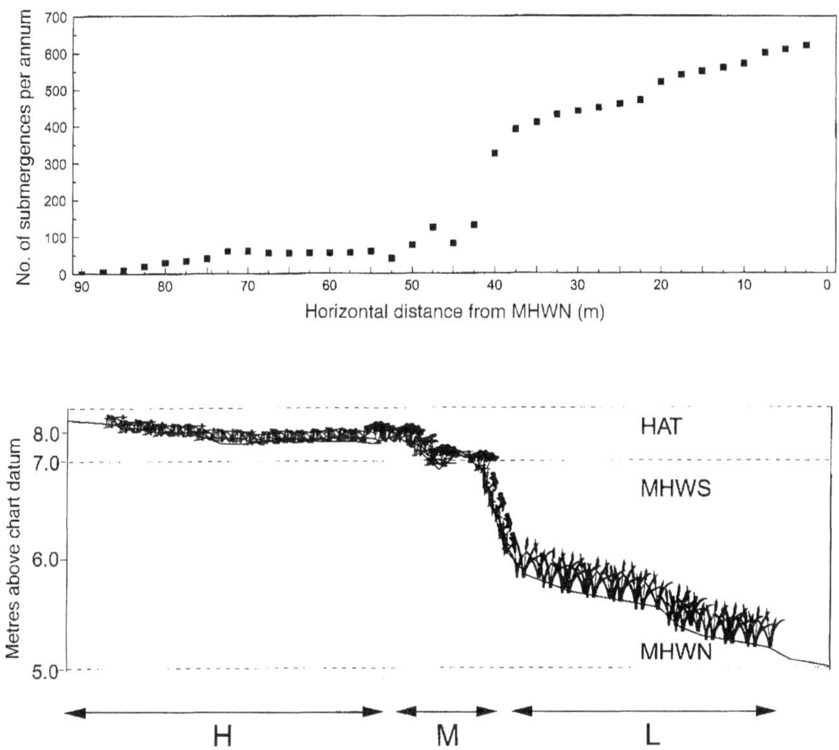

Figure 9.1 Profile of the saltmarsh at Brentlass in Milford Haven, south-west Wales. Low marsh (L) dominated by *Spartina anglica*; middle marsh (M) dominated by *Puccinellia maritima* or mixed communities including *Armeria maritima* and *Plantago maritima*; and high marsh (H) dominated by *Festuca rubra* (data from Dalby, 1970). The number of submergences per annum for points along the transect are shown in graph above profile. After Long and Mason (1983).

major concern (Burd, 1992). Harmsworth and Long (1986) estimate that saltmarsh at Dengie, Essex, has declined by approximately 10% between 1960 and 1981 at rates of between −16.0 and −1.0 ha/year. The land being lost is predominantly from areas of *Puccinellia maritima* (common saltmarsh-grass) and *Atriplex portulacoides* (= *Halimione portulacoides*, sea-purslane), i.e. middle-marsh communities – not, as might be expected, the pioneer zones of the marsh. In places, the eroding seaward edge of the marsh is marked by a small earth cliff, the base of which is easily undercut by the scouring action of seawater, causing a sudden and substantial decrease in the height of the marsh and reducing the likelihood of recolonization. It has been suggested that if this processes continues the marsh would be effectively lost within 50 years. Although, as appears to be the case here, saltmarsh cliffs are frequently associated with erosion

(Long and Mason, 1983), this is not always so. Sediment eroded from cliffs may be reworked and accreted within the marsh. A stepped profile may be indicative of maturity; where the seaward development of the marsh has been reached, a stepped profile may develop if accretion continues over the vegetated portion of the marsh (Carter, 1988).

More recent work by Burd (1992) has confirmed the findings Harmsworth and Long (1986). The observed rates of marsh loss at Dengie are not exceptional, the majority of saltmarshes in the greater Thames having suffered substantial losses between 1973 and 1983 (Burd, 1992). The reasons for these losses are complex and not fully understood. Undoubtedly, one major contributing factor is the sinking of the east coast (Harmsworth and Long, 1986; Burd, 1989, 1992). The British Isles are tilting along a south-west/north-east axis, causing a rise in sea levels relative to the land surface in the south-east (Pirazzoli, 1986). The Essex coast is estimated to be sinking at a rate of 3 mm/year (Burd, 1992). On sinking coastlines, saltmarshes may continue to extend seaward as long as conditions and sediment supply allow the accretion rate to exceed, and thus compensate for, the rate of sinking (Ranwell, 1972; Long and Mason, 1983). The impact of the sinking coastline will be compounded by any global rises in sea levels (Burd, 1992). The die-back of *Zostera* beds during the early 1930s may also have had an impact on marsh erosion. By reducing wave energy, offshore *Zostera* beds favour the seaward development of the marsh. *Zostera* beds are particularly sensitive to changes in the rate of sedimentation and the elevation of mudflats (Boorman and Ranwell, 1977), and will therefore also be adversely affected by changes in the sea level relative to land.

The impact of the sinking coastline is exacerbated by the presence of sea defences. Sea-walls truncate saltmarsh development, preventing the landward migration of vegetation in response to rising sea levels. Harmsworth and Long (1986) also point out that sea walls may accelerate erosion by altering tidal flows and are characterized on many Essex marshes by an unvegetated eroded zone at their base.

9.3 INTERTIDAL SEAGRASS COMMUNITIES

English Channel and North Sea coast seagrass (or eelgrass) communities contain only two angiosperm species: *Zostera noltii* (dwarf eelgrass) and *Z. marina* (eelgrass). The narrow-leaved form of *Z. marina*, variety *augustifloria*, found throughout the greater Thames estuary (Boorman and Ranwell, 1977), was recognized by Stace (1991) as a third species, *Z. angustifolia* (narrow-leaved eelgrass). Unlike the majority of angiosperms, these species are able to complete their life cyles whilst completely submerged in sea water. *Zostera noltii* is widely distributed throughout western Europe, frequently forming extensive intertidal belts below the mean high water at neap tides but not extending into the sublittoral zone (Den Hartog, 1983). *Zostera marina* is described by Den Hartog (1983) as an opportunist species, able to colonize a wide range of substrates from soft mud to almost gravelly sands and tolerant of low salinities. It occurs widely in shallow coastal waters from mean tide level to several metres below Chart Datum. Typically *Z. marina* occurs below *Z. noltii*.

In the greater Thames estuary, Wyer *et al.* (1977) reported that *Z. noltii* is normally associated with free draining hummocks, whilst *Z. marina* (narrow-leaf form) usually occupies hollows lower down the shore which remain filled with standing water at low tide. Therefore, where mixed stands occur, *Z. noltii* is largely confined to hummocks and *Z. marina* to wetter hollows.

Although perennial forms of *Z. marina* exist, an annual life history predominates in temperate climates. Intertidal *Zostera marina* is particularly sensitive to low temperatures and frost damage, dying off during September and establishing the following April from seed (Boorman and Ranwell, 1977; Wyer *et al.* 1977; Den Hartog, 1983). In contrast, *Z. noltii* is a wintergreen perennial, able to grow throughout the year at water temperatures of 5°C, or above.

Given their angiosperm composition, eelgrass beds appear to extremely simple communities. However, Den Hartog (1983) has identified 16 ecologically distinct groups of plants characteristic of temperate *Zostera* beds (e.g. epiphytes, algal films on bottom substrata, mats of loose entangled algae). The taxonomic composition of these groups is poorly defined, despite their undoubted ecological importance. The entangled algal mat accounts for a significant proportion of community above-ground biomass (10–20%) and productivity. In addition, *Zostera* beds will support unique meiofauna and macrofauna assemblages. Den Hartog (1983) recognized five basic types of eelgrass beds, one of which, dominated by *Z. noltii* and associated with detritus-rich fine sand sediments and in which annual forms of *Z. marina* may occur, is widely distributed along the south and east coasts of Britain.

Offshore *Zostera* beds offer protection to saltmarshes by damping wave action and reducing the likelihood of marsh erosion. In some sites *Zostera* beds can be considered as the ultimate seaward limit of saltmarshes, for which they may act as a precursor, being displaced as saltmarsh advances seaward. However, they frequently demonstrate a separate developmental cycle in which they are not colonized by saltmarsh vegetation. The wintergreen *Z. noltii* is capable of increasing accretion and consolidating accreted material, causing mudflats to increase in elevation. Since *Zostera* species are particularly sensitive to changes in mudflat elevation, this process can ultimately result in the loss of *Zostera* and erosion of the formerly stabilized mudflat, which may or may not be subsequently recolonized by *Zostera*. Transplant experiments suggest that eelgrass beds are only capable of becoming established and persisting on mudflats where the surface height varies by ±7 cm/year (Boorman and Ranwell, 1977). Many *Zostera* beds, particularly those in estuaries, are subjected to fluctuating environmental conditions, disturbance and disruption from storm events and are therefore essentially temporary systems, remaining in the pioneer stages of development (Chapman, 1976; Boorman and Ranwell, 1977; Den Hartog, 1983, 1987).

Substantial *Zostera* beds exist within the greater Thames estuary. Based on data collected in the early 1970s, Wyer *et al.* (1977) estimated the total area of *Zostera* to be 494.2 ha, the majority located at two sites: Marlin Sands (60.7%) and Leigh-on-Sea, Essex (19.3%). Relatively small areas occur in the Medway and Swale estuaries. Along the Essex coast the *Z. noltii* occupies an area (320.6 ha) approximately three times the area occupied by *Z. marina* (narrow-leaved form). However, in terms of standing biomass, *Z. marina* is approximately three times as productive per unit area. Freshweight standing crop estimates of 0.615 and 1.7 tonnes/ha have been obtained for *Z. marina* and *Z. noltii*, respectively. Standing crops vary throughout the year. During 1973 to 1974, the freshweight standing crop of *Z. noltii* increased from 20 g m^2 during January to May, to 63 g m^2 in July and 100 g m^2 in September, after which it decreased due to leaf loss during storm events, reduced growth and as a direct result of wildfowl grazing. In the case of *Z. marina* little standing biomass was evident until March (1.0 g m^2); this increased rapidly through the year, peaking at 410 g m^2 in September, after which die-back and rapid leaf loss occurred (Wyer *et al.* 1977).

The *Zostera* beds of the greater Thames are of major ecological importance, providing the principle winter grazing food for an estimated 20% of the world population of Brent geese (*Branta bernicla* L.), which preferentially graze *Z. noltii.* These beds are also exploited by several other wildfowl species notably widgeon (*Anas penelope* L.) (Charman, 1977). From the time of their arrival in September until January, Brent geese in Essex exploit the *Zostera* beds at Maplin and Leigh-on-Sea, after which they may be found feeding on green algae (e.g. *Enteromorpha* and *Ulva* spp.) in other estuaries. As food resources become scarce, the geese move on to neighbouring agricultural land. Prior to the early 1970s, feeding by Brent geese on agricultural land had only been recorded during the severe winter of 1962–1963 (Boorman and Ranwell, 1977). In most years food stocks are eaten out by early winter, causing a subsequent move to less preferred and less traditional food supplies (Charman, 1977). Individual geese consume some 121.6 g (dry weight) per day of *Z. noltii*; given this level of consumption, the *Zostera* beds at Maplin are thought to be capable of providing one million goose days of grazing (Boorman and Ranwell, 1977). In south east England, *Zostera* beds provide approximately 50% of the winter food requirements of migrating Brent geese (K. Charman, personal communication, cited in Wyer *et al.*, 1977).

The relationship between Brent geese populations and *Z. marina* is not clear. Since *Z. marina* is not wintergreen, with standing biomass decreasing rapidly prior to the arrival of the geese, it is unlikely to contribute significantly to their diet. However, anecdotal (e.g. old wildfowling) authorities persistently refer to Brent geese grazing on *Z. marina* and the birds have been noted feeding further out on lower mudflats where *Z. marina* might be expected to occur (Charman, 1977).

During the 1930s, European *Zostera* beds were devastated by a 'wasting disease', which brought *Z. marina* to the brink of extinction; *Z. noltii* was largely unaffected. The primary cause of the decline, from which *Z. marina* has subsequently recovered, remains unknown (Den Hartog, 1987). The rapid decline in *Z. marina* was parallelled by a 25% reduction in Brent geese from their pre-1930s levels. Charman (1977) suggested that the current preference for *Z. noltii* may represent an adaptive feeding response to a decline in the abundance of *Z. marina*. Although attractive, this suggestion is difficult to reconcile with the annual habit of *Z. marina*, unless conditions prevailing prior to 1930 allowed perennial forms of *Z. marina* to persist throughout the winter.

In addition to being directly exploited by grazing wildfowl, a substantial quantity of *Zostera* biomass enter the detritus food chain, helping to maintain mudflat productivity. Locally significant amounts of detritus may be washed up on neighbouring saltmarshes and beaches, creating suitable conditions for drift-line communities dominated by members of the Chenopodiaceae, particularly *Atriplex littoralis* (grass-leaved orache) and *Atriplex prostrata* (= *A. hastata*, spear-leaved orache) (Den Hartog, 1983).

The current extent of *Zostera* beds is uncertain. It is frequently asserted that they have not fully recovered from the wasting disease of the 1930s. One possible reason for this is the invasion and colonization of mudflats which previously supported *Zostera* by *Spartina anglica*. The situation is complicated by the lack of reliable recent survey data and the essentially ephemeral nature of the beds. Wyer *et al.* (1977) reported that from an initial area of approximately 48.21 ha in 1946, the *Zostera* beds at Leigh-on-Sea, Essex, declined to 8.37 ha in 1953. This decline has been attributed to mechanical damage and sedimentation following a period of severe coastal flooding. After 1953 the beds recovered, reaching a size of 84.94 ha in 1973.

9.4 SALTMARSH VASCULAR PLANTS AND COMMUNITIES

Some 325 species of vascular plants have been recorded on British saltmarshes, of which about 250 are relatively widely distributed. However, only some 45 species can be considered as halophytic and specifically restricted to saltmarshes. The exact number of these saltmarsh species depends partially on how species are defined (Adam, 1990). For example, in the taxonomically difficult genus of *Salicornia* (glassworts), at least 20–30 'sorts' can be distinguished in south-east England alone. Stace (1991) considers that they probably belong to only three species, *S. procumbens* agg., *S. europaea* agg. (common glasswort) and *S. pusilla* (one-flowered glasswort). *Salicornia europaea* agg. and *S. ramosissima* (purple glasswort) are reported as commonly occurring on the saltmarshes of the greater Thames. However, recent work suggests that *S. ramosissima* is not distinct from *S. europaea* agg. (Ingrouille and Pearson, 1987).

Following the work of Chapman (1941), east, west and south coast saltmarshes have been viewed as distinct types. Chapman's classification of British saltmarshes was based on an interpretation of the supposed successional development of saltmarshes. These successional patterns of development were largely deduced from the zonation of saltmarsh vegetation (Adam, 1978, 1990). In this scheme, east coast saltmarshes (Figure 9.2), including those of the greater Thames estuary, are characterized by the abundance of a middle-marsh community known as the general salt marsh (GSM) community (Chapman, 1934, 1964). The co-dominants of this community are: *Armeria maritima* (thrift), *Limonium vulgare* (common sea-lavender), *Plantago maritima* (sea plantain), *Spergularia media* (greater sea-spurrey) and *Triglochin maritimum* (sea arrowgrass). Within the GSM, these species frequently occur within a grass matrix of *Puccinellia maritima*. On well drained, slightly raised portions of the middle marsh, often associated with raised creek banks, *Atriplex portulacoides* may become locally abundant. At higher levels on the marsh, *Parapholis strigosa* (sea hard-grass) may also be associated with the community. On the south coast, the extent of the GSM community is severely restricted by the presence of *Spartina* (cord-grasses), whilst grazing pressure and sandy substratum are thought to limit its development on the grass dominated marshes of the west coast (Chapman, 1964).

Few marshes demonstrate a completely untruncated pattern of zonation. The landward limits, particularly of east coast marshes, are frequently set by the presence of sea defences and agricultural activity. Runoff from surrounding agricultural land can result in localized eutrophication, whilst freshwater inputs can influence the successional development of the upper marsh. This, coupled with the difficulty of interpreting plant zonation, limits the usefulness of Chapman's scheme. In addition, since Chapman's original work, the development and subsequent zonation of saltmarsh vegetation has undoubtedly been affected by the dramatic increase in the extent and distribution of *Spartina anglica* (common cord-grass). These problems have been compounded by the widespread inconsistent use and definition of the GSM community (Chapman, 1976; Adam, 1978).

9.5 THE ORIGIN AND SPREAD OF *SPARTINA*

Over the last century the spread of *Spartina* has fundamentally altered the ecology of northern European saltmarshes. Prior to 1870 two species

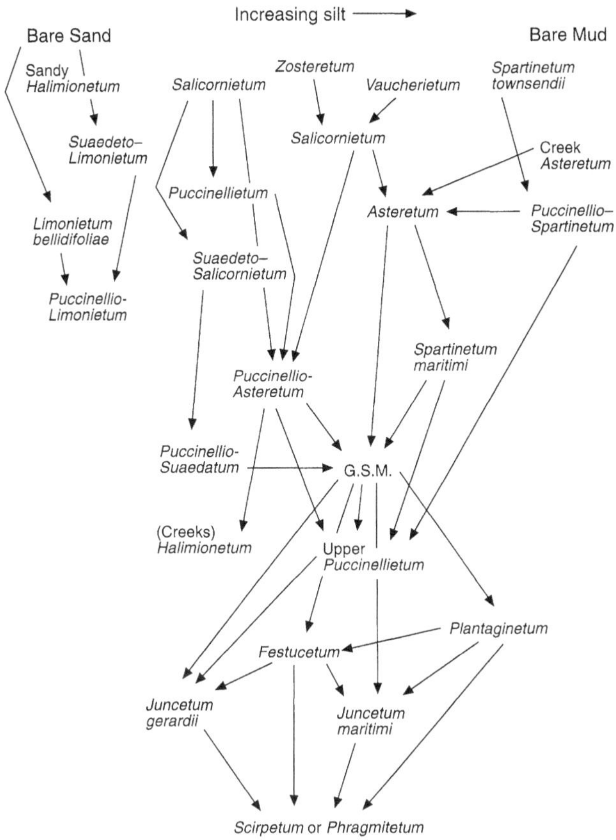

Figure 9.2 Idealized seral east coast development diagram. After Chapman (1964).

occurred in Britain. The native species, *Spartina maritima* (small cord-grass) was restricted to south-east England and was probably only ever locally a community dominant. The second species, *S. alterniflora* (smooth cord-grass), which is believed to have been introduced into Southampton Water in ships' ballast, was first recorded in 1824. By the turn of the century this species was widespread and locally abundant along the coasts of Hampshire. In 1870 a new form, *S.* × *townsendii* (Townsend's cord-grass), was recorded at Hythe in Southampton Water. This form spread rapidly and because of its vigour and ability to increase accretion; it was widely planted throughout Europe to stabilize mudflats, aid the reclamation of intertidal land and reduce coastal erosion. From these initially scattered introductions, *Spartina* spread rapidly along the coasts of Britain and northern Europe. *Spartina* × *townsendii* is thought to have arisen from the natural hybridization of *S. maritima* and *S. alterniflora*. The original forms of *S.* × *townsendii* were sterile; fertile forms were not collected until 1892. The fertile form, now know as *S. anglica*, was probably derived from *S.* × *townsendii* via chromosome doubling (Goodman *et al.*, 1959; Marchant, 1968), the two species being morphologically very similar. Because of the difficulty in distinguishing between the two species, it is probable that the majority of areas previously thought to have been planted with *S.* × *townsendii* were actually planted with the fertile *S. anglica*, the spread of which has been largely responsible for the rapid expansion of *Spartina* (Adam, 1990).

Today, *Spartina anglica* is still spreading and colonizing new sites, especially in the north and west of England. In contrast, *Spartina maritima*, *S. alterniflora* and *S.* × *townsendii* now occupy only remnants of their former actual and supposed distributions. *Spartina alterniflora* is thought to be restricted to one site in south Hampshire, *S.* × *townsendii* occurs locally from Dorset to West Sussex, although it may be found scattered elsewhere in southern Britain, and *S. maritima* occurs only locally in south and east England from the Isle of Wight through to North Lincolnshire (Adam, 1990; Stace, 1991).

The decline of both *S. maritima* and *S. alterniflora* is, in part, due to their displacement by the more vigorous and competitive *S. anglica*, although in the case of *S. maritima* climatic factors may also be important. This species is extremely rare throughout northern Europe, where it displays low vigour and partial seed sterility. Plants of *S. maritima* from southern Europe, towards the centre of its distribution, display considerably more vigour. In Britain the species is unable to tolerate high accretion rates and is largely confined to pits and depressions on the upper marsh in south-east England. Extensive low-marsh stands of *S. maritima* occur only at Maplin Sands, Essex (Boorman and Ranwell, 1977; Adam, 1978, 1990).

Because *S. anglica* is more tolerant of tidal submergence than any other saltmarsh species, it has been able to colonize mudflats which previously would not sustain saltmarsh development. On many marshes, stands of *S. anglica* have largely replaced the pioneer *Salicornia* (glasswort) community and locally it has invaded and displaced eelgrass communities. Rather than competitively displacing the *Zostera* beds, it is possible that *S. anglica* has simply colonized sites vacated by *Zostera* following the wasting disease of the early 1930s. However, the continued presence of *S. anglica* will prevent the re-establishment of the *Zostera* beds.

Given the high productivity of *S. anglica* (Long and Woolhouse, 1979; Hussey and Long, 1982), it is likely that the productivity of many marshes has increased as *S. anglica* has become established. The consequences of this, and the impact of the resulting litter on saltmarsh development and the ecology of neighbouring mudflats, is largely unknown. Beeftink (1975) suggested that the amount of litter produced, and its uniform composition, reduced the diversity of strandline vegetation in Holland. Of concern also is the effect that the seaward development of *S. anglica* may have on the subsequent development of the marsh profile. The ability of *S. anglica* to build and accrete fine silt will affect the distribution of sediments over the marsh surface. On marshes with well developed lower zones dominated by *S. anglica*, the upper marsh may become starved of fine sediment. As the lower marsh continues to build, a convex marsh profile may develop, inhibiting drainage and runoff from the mid and upper marsh. Such changes to marsh hydrology would have a major impact on the vegetation and future development of the marsh (Ranwell, 1972; Adam, 1990).

Although *S. anglica* has spread rapidly, displacing previous pioneer communities, it has failed to become widely established at higher elevations amongst the closed communities of the mid marsh (Beeftink, 1977; Adam, 1990). It appears to be sensitive to trampling and grazing pressures, which inhibits the development of a closed *Spartina* sward, allowing the establishment of *Puccinellia maritima*. Low-marsh stands of *S. anglica* have been subject to localized die-back. In some sites (e.g. Poole Harbour, Dorset) the extent of die-back has been substantial. Although *S. anglica* dieback is most marked along the Channel coasts (Burd, 1989; Adam, 1990), it is also evident at other sites – for example, Bridgewater Bay, Somerset, where in places *S. anglica* along the seaward edge of the marsh has been substantially eroded. Die-back may be only a temporary phenomenon; unpublished records held by English Nature describe *S. anglica* at Chichester Harbour, West Sussex, as being 'moribund' in 1981 and as growing vigorously in 1984. The causes of die-back are unknown and there is no evidence of primary pathogen involvement. Local impedance of drainage, the development of toxic soil conditions, nutrient depletion and the accumulation of toxic metals from anthropogenic sources have all been suggested as possible causes. As Adam (1990) states, it is still too early to assess the full impact of *S. anglica*. Because of the relatively recent origin of *S. anglica* it is unsafe to assume, as Chapman's serial diagrams suggest (Figure 9.2), that *Spartina*-dominated lower-marsh communities will be replaced by the same communities that currently occupy older and higher parts of the marsh. Given the ecological importance of both saltmarshes and intertidal mudflats, it is essential that the long-term effects of *Spartina* on these habitats be established.

9.6 REGIONAL SALTMARSH COMMUNITIES

An objective classification of British saltmarsh vegetation, using conventional phytosociological table rearrangement procedures (Mueller-Dombois and Ellenberg, 1974), is presented by Adam (1978, 1981). Based on an analysis of the floristic composition of some 3000 samples (normally 2 m × 2 m) taken largely from the middle and lower-marsh areas at 133 sites, Adam (1976, 1978) was able to identify 60 more or less well defined

Table 9.1 Selected relevant key distinguishing species, characteristic of saltmarsh nodum groups identified by Adam (1978)

Species	*Nodum group*							
Spartina anglica	I							
Salicornia spp	I							
Aster tripolium		II						
var. *discoideus*		II						
Atriplex portulacoides		II						
Limonium vulgare		II						
Armeria maritima		II						
Atriplex littoralis			III					
Bolboschoenus maritimus (*Scirpus maritimus*)			III					
Phragmites australis			III					
Juncus gerardii			III					
Elymus repens (*Agropyron repens*)			III					
Glaux maritima			III					
Leontodon autumnalis				IV				
Oenanthe lachenalii				IV				
Agrostis stolonifera				IV	V			
Trifolium repens					V			
Suaeda vera						VIII		
Limonium binervosum						VIII		
Frankenia laevis						VIII		
Filipendula ulmaria							IX	
Carex flacca							IX	
Iris pseudacorus							IX	
Spartina maritima								X
Festuca rubra		II	III	IV	V		IX	
Plantago maritima		II	III	IV	V		IX	
Puccinellia maritima	I	II	III		V	VIII	IX	

vegetation noda or species associations. Taking into account the frequency and geographical distribution of derived noda, Adam (1981) presented a dichotomous key to British saltmarsh communities, identifying 49 distinct principal vegetational units (noda) or communities.

By undertaking a two-way classification using numerical cluster routines to group related noda together into major nodum groups and then classifying the saltmarsh sites on the basis of the occurrence of these nodum groups, the data may be summarized in a single-ordered two-way table containing 10 fundamental nodum groups and 13 site groups distinguishable in terms of their vegetation (Adam, 1978). Three fundamental saltmarsh types can be distinguished, differing largely on the occurrence of vegetation within the nodum groups I, II, IV and V (Table 9.1, Figure 9.3).

Type A marshes occur predominantly in south-east England. They are characterized by the high frequency of plant communities associated with the low marsh, pioneer habitats (Nodum group I) and the extent of low to

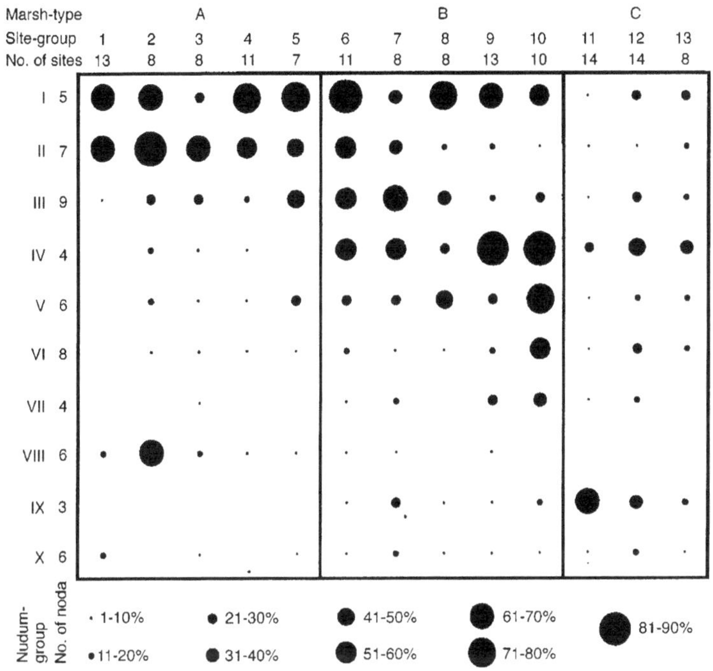

Figure 9.3 Summary two-way table of site-vegetation groups. Proportional representation of 10 nodum groups in the 13 site groups indicated by circle size. Taken from Adam (1978).

middle communities distinguished by the abundance of *Armeria maritima, Atriplex portulacoides, Limonium vulgare,* the rayless form of sea aster (*Aster tripolium* var. *discoideus*) and *Puccinellia maritima.*

Type B marshes occur on the west coast and are characterized by an increase in the occurrence of the noda in groups III, IV and V. Nodums IV and V consist largely of middle to upper-marsh grass sward communities of *Festuca rubra* (red fescue) or *Agrostis stolonifera* (creeping bent), in which glycophyte species such as *Trifolium repens* (white clover) and *Leontodon autumnalis* (autumn hawkbit) are abundant. Nodum III, which is largely restricted to wet upper-marsh areas, contains the more salt-tolerant species *Atriplex littoralis* and *Bolboschoenus maritimus* (= *Scirpus maritimus,* sea club-rush), along with glycophytes such as *Phragmites australis* (common reed).

All Type C marshes occur in Scotland and are characterized by the relative scarcity of plant groups evident on type A and B marshes.

Greater Thames marshes are all of Type A. Those of Essex (i.e. sites in group I) are clearly distinguished from other group A marshes by Nodum

Table 9.2 Description of saltmarsh survey communities types used by Burd (1989) and their relationship with the proposed NVC class

Saltmarsh types	*NVC class*	*Notes*
1. *Spartina*	SM4 *Spartina maritima* SM5 *S. alterniflora* SM6 *S. anglica*	*Spartina*-dominated pioneer community, usually monospecific
2. *Salicornia/Suaeda*	SM7 *Arthrocnemum perenne* SM8 Annual *Salicornia* SM9 *Suaeda maritima*	Open pioneer community, seaward edge of marsh, or in depressions or pans
3. *Aster*	SM11 *Aster tripolium* var. *discoideus* SM12 Rayed *Aster tripolium*	Often occurs as monospecific stands usually above 2, or along creek side
4. *Puccinellia*	SM10 Transitional low-marsh vegetation SM11 *Puccinellia maritima*	Low–mid marsh, *Puccinellia* intermingled with several other species, e.g. *Aster* or *Limonium*
5. *Halimione*	SM14 *Halimione protulacoides*	Low–mid marsh, may occur as pure stands, intermingled with 4. *Artemisia maritima* often found in association
6. *Limonium/Armeria*	SM13 *Puccinellia maritima–Limonium/Armeria* subcommunity	Diverse mid–upper-marsh community
7. *Puccinellia/Festuca*	SM13 *Puccinellia maritima–Gluax maritima; Plantago/Armeria* and turf fucoid subcommunities	With reduction in tidal inundation, *Festuca rubra* becomes more abundant
8. *Junus gerardii*	SM16 *Festuca rubra*	Upper-marsh community
9. *Juncus maritimus*	SM15 *Juncus maritimus/Triglochin maritimus* SM18 *Juncus maritimus*	Upper-marsh community
10. *Agropyron (Elymus)*	SM24 *Elymus pycnanthus* SM28 *Elymus repens*	Drift-line community
11. *Suaeda fruitcosa*	SM25 *Suaeda vera* drift community	Forms bushy stands along drift line
12. *Scirpus maritimus*	S21 *Scirpus maritimus*	Upper-marsh swamp communities
13. *Phragmites australis*	S4 *Phragmites australis*	
14. *Typha latiflora*	S19 *Eleocharis palustris*	
15. *Schoenoplectus tabernaemontani*	S20 *Scirpus lacustris*	
16. Shingle/dune	SM21 *Suaeda vera/Limonium binervosum*	Shingle/dune transition community
17. Freshwater transitions	MG11 *F. rubra/A. stolonifera*	Freshwater transition community

Note: additional saltmarsh groups are defined by Burd (1989) but do not often occur on the saltmarshes of the greater Thames

X, containing *Spartina maritima*. Similarly the North Norfolk marshes may be identified by abundance of Nodum VIII, a sand dune transition community characterized by the presence of *Suaeda vera* (*S. fruticosa*, shrubby sea-blite).

Figure 9.4 Geographical saltmarsh units identified by Burd (1989): **1**. East and South-East England; **2**. South-West England; **3**. Wales; **4**. North-West England and South-West Scotland; **5**. West Scotland; **6**. East Scotland. After Burd (1989).

An extensive survey of British saltmarshes has been published by English Nature (Burd, 1989). Including upper-marsh transition communities, the survey recognized 17 principal saltmarsh community or vegetation types. These, and their relationship to the draft saltmarsh National Vegetation Classification scheme, are outlined in Table 9.2.

The presence and area of each vegetation type was recorded for virtually every significant area of saltmarsh. Whilst the results of the survey are broadly similar to those of Adam (1978), six distinctive regional salt marsh types were identified (Figure 9.4). In many respects these may been seen as regional variants of the three fundamental marsh types described by Adam (1978). The saltmarshes of the south-east are characterized by the scarcity of transitional and upper marsh swamp communities (types 12 and 13) and the virtual absence of communities characterized by *Juncus gerardii* (saltmarsh rush) (type 8) and *J. maritimus*

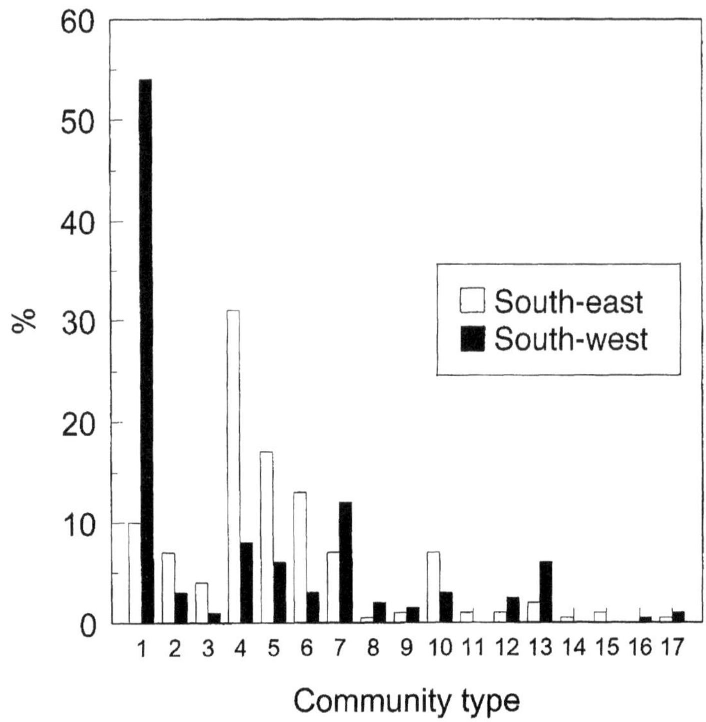

Figure 9.5 Community composition of south-east and south-west saltmarshes. Percentage representation of saltmarsh community types (1 to 17). Outline of community types given in Table 9.2.

(sea rush) (type 9) which are common on the upper grazed marshes of the west. In contrast the low–mid *Puccinellia maritima* (4), sea purslane (*Halimione*) (5) and mid—upper sea lavender and thrift (*Limonium–Armeria*) (6) community types are particularly abundant (Figure 9.5). The drift-line community, characterized by the presence of *S. vera*, has the most geographically restricted range, being recorded only along the north Norfolk coast (Burd, 1989). This probably represents the northernmost limit of its distribution in Britain (Chapman, 1976).

From the appendices of the English Nature report it is possible to extract data on the saltmarsh vegetation of individual counties. Approximately 78% of the combined records for Essex and the north Kent coast (some 5492 ha) occur within the greater Thames estuary. In terms of their community composition, the saltmarshes of Essex and the north Kent coast are virtually identical, supporting the assertion by Adam (1978). The community composition of these greater Thames and the south coast marshes (obtained by combining data for the saltmarshes of Hampshire

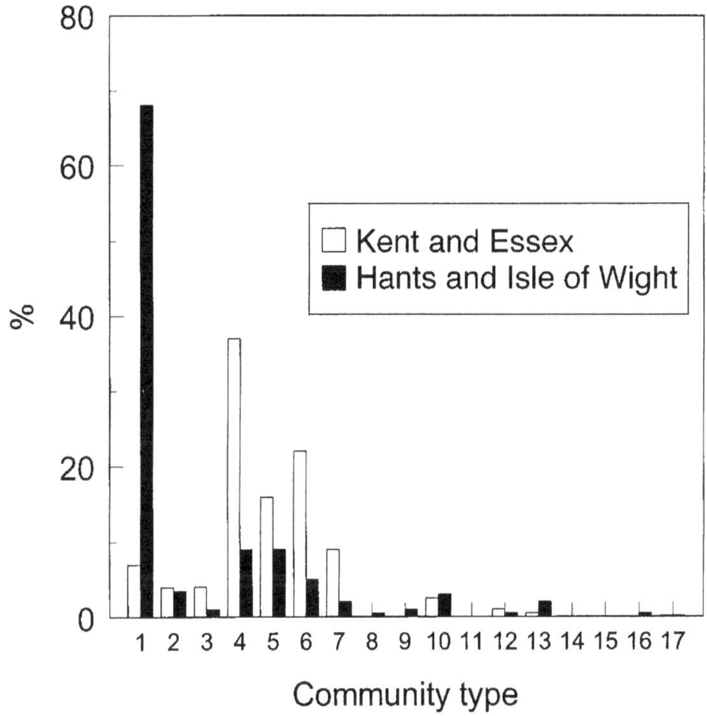

Figure 9.6 Comparison of the community composition of the greater Thames and south coast marshes. Percentage representation of saltmarsh community types (1 to 17). Outline of community types given in Table 9.2.

and the Isle of Wight) is presented in Figure 9.6. Major differences between these two groups of marshes are readily evident. The mid-marsh zones of the greater Thames support substantial areas of communities dominated by *Puccinellia maritima* (type 4), sea purslane (*Halimone*) (type 5), the sea lavender–thrift association (*Limonium–Armeria*) (type 6) and to a lesser extent the mid- to upper-marsh *Puccinellia–Festuca* grass community type 7. In contrast, the south coast marshes are effectively dominated by the *Spartina* pioneer community (type 1), which accounts for approximately 68.5% of the 1762 ha surveyed.

The greater Thames saltmarshes are also clearly distinct from those of Norfolk (Figure 9.7). In comparison to Norfolk, greater Thames saltmarshes have extensive areas dominated by the mid–upper community types 4 (*Puccinellia*), 6 (*Limonium–Armeria*) and to a lesser extent 7 (*Puccinellia–Festuca*). The Norfolk marshes are characterized by the relative abundance of the *Salicornia–Suaeda* pioneer community (type 2), the middle-marsh *Halimione* community type (5) and the two drift-line community types 10 (*Agropyron–Elymus*) and 11 (*Suaeda fruticosa*).

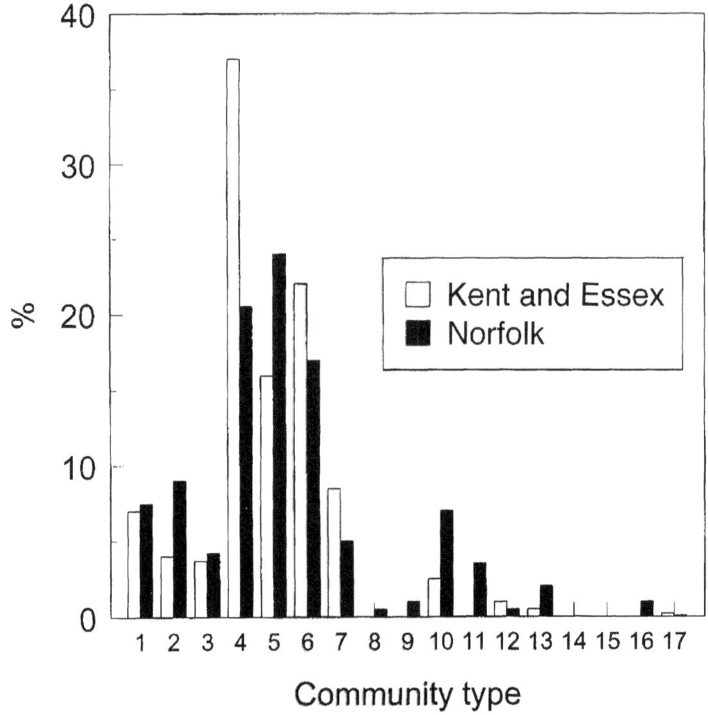

Figure 9.7 Comparison of the community composition of the greater Thames and Norfolk saltmarshes. Percentage representation of saltmarsh community types (1 to 17). Outline of community types given in Table 9.2.

9.7 SALTMARSH VEGETATION OF THE GREATER THAMES

Above the *Zostera* mudflats (where they occur), the pioneer lower marsh communities are typically dominated by stands of *Spartina anglica*, which are frequently pure. Where *S. anglica* is well established, *Salicornia europaea* agg. tends to be restricted to sites where a thin layer of gravel or shingle overlays hard clay (Adam, 1981). Occasionally, a small transitional band of vegetation with *S. europaea* agg. and *Puccinellia maritima* as co-dominants may persist above *S. anglica*. In the absence of *S. anglica*, the pioneer zone would be dominated by this association, the abundance of *P. maritima* increasing with marsh elevation.

The presence of *S. anglica* will dramatically affect rates of lower marsh accretion. In the absence of *S. anglica*, accretion is largely dependent on the presence of *P. maritima*, as *S. europaea* agg. is an annual and has little effect on net accretion. Lower-marsh communities dominated by *P. maritima* are not uncommon.

Abundant on the lower marsh, often growing in association with *P. maritima*, glasswort and *Suaeda maritima* (annual sea-blite), is the rayless form of sea aster (*Aster tripolium* var. *discoideus*). This association is particularly frequent and well developed on the saltmarshes of the Thames estuary and coasts of Essex and Kent, forming a distinctive zone either above *Spartina* or occasionally occurring as the lowest vegetation type (Myers, 1954; Fojt, 1978; Adam, 1978, 1981). The rayless form of sea aster tends to predominate on firm clays, or on low organic silts with a high content of shell fragments. Within the low marsh it is frequently absent from small depressions where *P. maritima*, *Suaeda maritima* and *S. europaea* may become locally dominant.

The occurrence and distribution of the rayed and rayless forms of *Aster tripolium* have been discussed by Dalby (1970) and Gray (1971, 1974). The development of ray florets appears to be governed by a single gene in a simple Mendelian manner. However, rayless forms differ from the rayed form of aster in several respects, e.g. age at time of first flowering, time of flowering, fruit weight and germination requirements, suggesting that a complex of genes linked to that governing floret development maybe involved (Gray, 1971). The rayed and non-rayed forms of *A. tripolium* also differ in their ecological distribution, the rayless form occurring at lower-marsh levels, whilst the rayed form tends to occur scattered throughout the middle and upper marsh (Myer, 1954; Dalby 1970; Gray, 1971). Myers (1954) reported that the rayed form was restricted to low salinity marsh sites, inland of the sea-wall at Leigh-on-Sea, Essex, whilst the rayless form only occurred seaward of the sea-wall on high salinity, 'natural saltmarsh' sites subjected to frequent flooding.

Unique to the greater Thames marshes is the occurrence of the now rare *Spartina maritima*, which can be found growing in association with *P. maritima*, *Salicornia europaea* agg., *Limonium vulgare* and in some cases *Suaeda maritima* and *Aster tripolium* (Boorman and Ranwell, 1977; Adam 1978, 1990). This community never occupies extensive areas, being restricted to shallow, soft-mud depressions in the middle marsh (Adam, 1981). Although *S. anglica* and *S. europaea* agg. are normally associated with the pioneer zone, they may also occur scattered throughout the marsh associated with small surface depressions, regions of impeded drainage and 'secondary marshes' developing on the remains of collapsed creek banks. Extensive low-marsh stands of *S. maritima* now occur only at Maplin Sands (Boorman and Ranwell, 1977; Adam 1978, 1981).

Atriplex portulacoides is dominant over extensive areas of the middle marsh, growing in association with *P. maritima*, *Limonium vulgare*, *Armeria maritima*, *Plantago maritima* (sea plantain) and *Triglochin maritimum* (Long and Mason, 1983). The saltmarsh rarity *Inula crithmoides* (golden-samphire) is a feature of the greater Thames marshes. Although *I. crithmoides* may be found growing in a wide range of communities, it tends to be particularly abundant in areas dominated by *A. portulacoides* (Boorman and Ranwell,

1977; Adam, 1978, 1981, 1990). The species is particularly abundant on saltmarshes with a high lime content; the most extensive populations are associated with the upper marshes of the Medway (Boorman and Ranwell, 1977).

Inula crithmoides and several other species, including *Frankenia laevis* (sea-heath), *Suaeda vera*, *Limonium bellidifolium* (matted sea-lavender), *L. binerivosum* (rock sea-lavender), *Spartina maritima*, *Sarcocornia perennis* (*Salicornia perennis*, perennial glasswort) and *Aster tripolium* var. *discoideus*, are locally abundant on south-eastern marshes but do not occur on marshes north of a line running roughly from the Wash to the Bristol Channel (Adam, 1978). Also within this group is *Atriplex pedunculata* (*Halimione pedunculata*, pedunculate sea-purslane) which was thought to be extinct, but was recently rediscovered in Essex. This is listed as an endangered European species. Its rediscovery was the first ever record in Essex and the first record in Britain for 50 years (Leach, 1988). Leach speculated on whether the record in Essex represented a long-standing population, or the recent establishment of seeds brought to the site from Denmark by Brent geese.

Middle-marsh vegetation is typically non-uniform in composition, and the occurrence of salt-pans, small variation in microtopography, sediment type, soil aeration and salinity are reflected in the distribution of species. For example, within an area 50 m by 10 m at Colne Point, Essex, in which the vertical elevation varied by less than 5 cm, Othman (1980) was able to identify six distinct groupings of species. Along a creek, fringe groups characterized by the presence of *A. portulacoides* occurred. Twenty metres from the creek these groups were replaced by groupings containing *L. vulgare*, *A. maritima*, *T. maritima*, *Plantago maritima* and *Spergularia marina* (lesser sea-spurrey). Furthest from the creek, on poorly drained, low elevation soils subject to large variations in salinity, *S. anglica* and *S. europaea* occurred.

At higher elevations the abundance of *A. portulacoides* and other associated halophytes declines and the representation of glycophyte species increases. *Puccinellia maritima* is frequently replaced by *Festuca rubra*. Approaching the drift line a diverse range of communities may occur, frequently dominated by *Elytrigia atherica* (*Agropyron pungens*, sea couch), with *A. prostrata* and *Juncus maritimus* also frequent, the exact composition often influenced by that of the neighbouring inland vegetation.

Unlike northern and western saltmarshes, transition to freshwater wetland habitats is rare. This partly reflects the frequent truncation of east coast marshes to the landward side, but may also result from the lower annual rainfall. Regional rainfall patterns may also partially explain the scarcity of glycophyte species associated with the upper marsh and the general lower diversity of south-east saltmarshes. However, west and south-east marshes differ in many other respects, including elevation and tidal range, zonation, sediment type and grazing. The possible roles of these factors in determining marsh vegetation has been extensively discussed by Adam (1978, 1981, 1990), who concludes that domestic grazing alone cannot explain the differences between west and south-east saltmarshes and suggests that substratum is an important factor. Given the extent and size of wildfowl population that exploits saltmarshes and neighbouring mudflats, wildfowl grazing activity may be another major factor, along with the extent of *S. anglica* invasion and establishment. Local and regional variations in saltmarsh vegetation appear to result from a complex and as yet poorly understood interaction between local (e.g. site history, previous and present grazing) and regional environmental factors (e.g. tidal range, sediment nature and climate).

The saltmarshes and associated *Zostera* beds of the greater Thames represent a unique and valuable natural resource of national and international importance, supporting substantial populations of wildfowl and wading birds. The vegetation of the greater Thames saltmarshes is clearly distinguishable from that of other coastal marshes in Britain and south-east England, and provides valuable sites and refugia for several rare plants. It is perhaps ironic that despite their proximity to major centres of human industrial and agricultural activity, these marshes appear to be most threatened by the consequences of natural geological processes.

CHAPTER 10

Estuaries: towards the next century

Alasdair McIntyre

10.1 INTRODUCTION

Estuaries are of unique importance. They were colonized by humans from the earliest times, having always been recognized as the link between land and sea, providing safe harbours and attractive sites for commerce and industry, commercial fishery resources, habitats for wildlife and opportunities for tourism and recreation. They are highly dynamic areas, subject to cycles of erosion and deposition and encompassing a complex of subtidal, intertidal and terrestrial habitats. The previous chapters have looked in some detail at one particular estuary: the Thames. This concluding chapter, still in the context of estuaries, looks ahead in a more general way towards the rapidly approaching new century. But first it is appropriate to glance backwards and briefly consider the previous 100 years. Such an exercise is always popular as we approach the end of a period, and there is no doubt that those reviewing the twentieth century will find an unusually rich mine to exploit, given the accelerating pace of technology, the increasing population growth, the rate of political change and the associated alterations in human attitudes and behaviour.

In the context of the marine environment the historian may well note that major changes have taken place particularly in the second half of the period. The 1950s brought radioactive fallout over all the oceans from nuclear weapon testing in the atmosphere and, alerted by the Minamata incident, an awareness of metal contamination in the sea from industrial effluents. With the 1960s came the damage from excessive use of pesticides in the form of synthetic organic compounds (e.g. DDT), and the threat of oil pollution from tanker wrecks. In the following years additional concerns developed. Discharges of sewage degraded beach amenities and

A Rehabilitated Estuarine Ecosystem. Edited by Martin J. Attrill.
Published in 1998 by Kluwer Academic Publishers, London. ISBN 0 412 49680 1.

posed a variety of public health problems, whilst increasing inputs of nutrients from agriculture, intensive stock-rearing and industry to poorly flushed marine areas generated excessive algae blooms, with various adverse effects, including deoxygenation of the water and associated kills of fish and even humans.

Shallow waters and coastal margins are particularly at risk, and warning signals for estuaries have been with us for some time. As previous chapters show, these warning signals for the Thames come not from the seaward side but from the much more vulnerable freshwater end. Indeed it has been said that the most important function of the Thames in the past was the carriage of sewage effluents, and more recently it has been suggested that many estuaries are almost the functional equivalents of industrial holding ponds on a continental scale. For the Thames we know that the water quality of the river declined rapidly in the last century due to sewage pollution, and that by 1856 the fisheries of the estuary had largely died out. It is almost a cliché to recall that the odour from the river at Westminster was once so bad that sittings of some committees of the House of Commons were interrupted, and that this went a long way towards stimulating appropriate legislation.

However, even at that time the situation was not new. Another well documented story tells that as far back as the year 1290, the stench from the river Fleet, a long-extinct tributary of the Thames, was, in the words of an old report, 'impossible to deaden with the strongest incense', and even to have caused the deaths of some of the brotherhood of White Friars who lived by the river (Wheeler, 1979). Some centuries later the river Fleet was built over and became simply a sewer. The river gave its name to the present Fleet Street in London, and one occasionally wonders if the street's later activities retained some of the river's earlier characteristics.

These points are made simply to emphasize that estuaries have always been at risk from their freshwater inflow. In the 1960s, estuaries of the American east coast were being referred to as 'the septic tanks of the megalopolis', and if such a statement could be made at that time about the most highly developed country in the world, how much greater must be the problems in the over-populated areas of the developing world. In the early 1980s, the Ganges river was described as 'gradually turning from river to drain, where raw sewage, garbage and muck pass continuously into the river from 361 outfalls' (Gourlay, 1992). The situation is similar elsewhere in Asia, the Caribbean and other parts of the world. But of course there are many more diverse problems located directly in the main bodies of the estuaries themselves, associated with the human activities focused there, and there is no need to go into detail on:

- the use of estuaries by shipping;
- their attractions to industry through access to the hinterland;
- their provision of cooling water for industrial installations;

- their potential as sites for barrages for power generation and water management;
- the opportunities they offer for a diversity of recreational pursuits;
- their use for receiving and dispersing wastes.

We need to recognize only that all these activities have significant consequences for the environment, not just in terms of polluting discharges from which recovery is possible, but also by the modification or elimination of habitats by land claim, draining and construction, which can lead to irreversible change. Land claim has affected at least 85% of British estuaries and has removed a substantial part of the intertidal area – in some estuaries over 80%. In the past 200 years take of estuaral land has been from 0.2 to 0.7% annually. Much of the land claimed is used for rubbish and spoil disposal, transport schemes, housing, car parks and marinas. Whilst all this may be justifiable in economic and sociological terms, it does constitute an irreplaceable loss of natural habitat. An additional problem is the perceived need to protect the coast from erosion and flooding. As much as 32% of the coast of England is protected by sea banks, walls, groynes and other artificial structures (Davidson *et al.*, 1991). Enclosure by linear sea defences results in direct loss of tidal land.

Another major concern is the building of barrages across the mouth of estuaries. These were probably first thought of as water storage schemes, but more recently for storm surge control (a noted example being in the Thames), tidal powers generation (with structures suggested for the Severn, Mersey, Wyre and Conway estuaries), and even recreational and urban renewal developments (e.g. the Taff in Cardiff). The latter are to some extent cosmetic, covering tidal mud flats and creating lakes for amenities and visual impact.

It is not unreasonable to suggest that of all marine environments, estuaries are likely to be the most vulnerable, if only because of the diversity of pressures on them. And as the world population continues to rise, and to show a net migration towards the coast, these pressures will intensify. It is worth looking at our present knowledge of estuaries, and how that has developed.

10.2 EARLY WORK ON ESTUARIES

A recent review (Davidson *et al.*, 1991) identified 155 British estuaries, occupying about 50% of our coastline and classified into nine separate groups: bar-built (47); coastal-plain (35); fjards (20); rias (15); embayments (13); complex (10); linear-shore (7); fjords (6); barrier-beach (2).

The very fact that such a confident classification could be made today is a significant advance. It is based on a great deal of earlier work worldwide aimed at assembling comprehensive data on estuaries. The biological focus at first tended to be on the types of plants and animals that lived there, and on physiological studies investigating how they were able to adapt to the rigorous conditions imposed by the frequent alterations of salinity, temperature and turbidity. Also, much effort was directed to defining estuaries, particularly in terms of their geomorphic and physical conditions.

These attempts at definition were not particularly conclusive, but at least the importance of estuaries was clearly appreciated, and in 1964 a wide range of interested parties got together at an international estuaries conference in Georgia (Lauff, 1967). Even at that time the question 'What is an estuary?' was well to the fore. The origins, the processes of formation, and the physical and biological characteristics of estuaries were discussed at length, yet the participants (all estuary researchers) still acknowledged the difficulty in agreeing to just what they were working on. Their definitions ranged from, on the one hand, restricting an estuary to the tidal stretch of a river to, on the other hand, expanding it to a major part of the contiguous ocean, on the argument that an estuary stretched out as far as a salmon at sea could detect the influence of its home run. The conference ended with a light-hearted attempt to define the ideal estuary. It was to have, among other things, tides, mixing of fresh and salt water, a well layered exchange system, an interesting variety of sediments, and, of crucial importance, it should be large enough to provide work for half a dozen researchers including one taxonomist, backed up by a dozen students, and there should be sufficient pollution to justify the budgets of the research workers! The present approach is rather less indulgent to scientists. However, that conference, in spite of much confusion, did demonstrate a good grasp of the basic characteristics of estuaries, and laid substantial groundwork. The years that followed were marked by a flow of meetings and seminars so that by the early 1970s scientists seemed to tire of the argument, agreed to differ on definitions, and turned to more fruitful research.

Over the next couple of decades this research covered a considerable diversity of interests, and a scan of some of the main journals devoted to estuaries in that period gives an indication of the topics of greatest interest. A very small number of papers dealt with structure, and since that was the nearest reference to the matter of definition, it seemed that argument had at last been laid to rest. The bulk of the contributions – more than a quarter of the total – focused on physical oceanography, covering tidal energetics, flushing rates, velocity profiles, turbulent mixing and temperature oscillations, as well as the transport, suspension and deposition of sediments, all clearly directed towards understanding the dynamics of estuarine processes.

Chemistry was also well to the fore, with two distinct areas of focus: one was on trace metals, their geochemistry, concentrations, distribution, speciation and fate; the other was on nutrients, with papers covering sources, transformations, regeneration and fluxes across interfaces. Turning to organisms, most attention amongst the plants was given to sea grasses and macroalgae, whilst for the animals the benthos occupied the top slot, followed by plankton.

On the topic of pollution, there were many studies of oil, while the effects of metals and the changes caused by excessive nutrients attracted interest. It is worth noting that the pollution papers came closer than any of the others to following through the story from sources and distribution to impacts and fate, thus integrating physical, chemical and biological aspects. It is also relevant that this all-inclusive approach is essential if the future management of estuaries is to be successful.

10.3 CONCERNS FOR THE FUTURE

Given the vulnerability of estuaries and the pressures on them discussed above, their adequate protection must be a major concern in the next few decades, and it is becoming increasingly recognized that conservation will need to involve more than just the traditional scientific aspects usually covered by the studies reported in earlier estuarine journals. These studies have, of course, laid the necessary foundations, but much more is required because of the diversity of interests within estuaries that can affect them, and of the range of hinterland activities that can impinge on them.

The extent to which inland events can have impacts on the coast is becoming increasingly apparent. It has been demonstrated, for example, that deforestation of mountainous regions 100 miles inland in the Philippines was responsible for the death of corals offshore by siltation. Again, in many parts of the world, the damming of rivers far from the sea has totally altered the hydrographic regime in estuaries. In Africa and North America at least 20% of the land runoff now originates from impoundments, and as more dams are built we must expect to see major alterations to coastlines on a large scale. In Nigeria, the flood plain of the Hadejia River has been reduced by over 300 km^2 as a result of dam construction (Dugan, 1990).

These, of course, are dramatic examples of the fact that far field effects must be taken into account, and that the protection of estuaries requires integrated management at the level of whole river systems. But the concept of integrated management must be applied also to activities within the estuary itself, a fact that was well spelled out recently by the UK's House of Commons Environment Committee in its Report on Coastal Zone Protection and Planning (HMSO, 1992). It pointed out that the coastline should not be seen as a physical or administrative boundary, but that the coastal zone should be treated as a unit embracing inshore waters, intertidal areas and maritime land. The Report also drew attention to the duplication of responsibilities and lack of coordination between the many organizations with interests in the coastal zone. Non-governmental organizations (NGOs), including the RSPB and the Marine Conservation Society, have noted the piecemeal nature of many coastal developments, each small in itself but adding up to a major removal of habitat, and have called for an integrated system of coastal planning and management, guided by an overall national strategy.

These ideas were strongly promoted at the Earth Summit Conference in Rio in 1992, and Chapter 17 of its Agenda 21 (which focuses on the protection of the oceans) advocates 'integrated management and sustainable development of coastal areas' (Grubb *et al.*, 1993). This has been taken up by, among other bodies, the Intergovernmental Oceanographic Commission (IOC) of UNESCO, which has established a programme of Integrated Coastal Area Management (ICAM).

Fortunately, all this has been pursued vigorously in the UK by conservation agencies and by NGOs. English Nature is engaged in a series of far-reaching management plans for estuaries to identify potential conflicts of interest and promote sustainable and multiple use, while the Scottish Focus on Firths programme provides a comparable approach – particularly in the Moray Firth and the Firth of Forth, where a forum for local interest groups has been established, community involvement strategies introduced and basic research projects initiated. On the wildlife protection side the RAMSAR Convention is becoming increasingly effective, and now there is the European Union's focus on conservation: the Habitats Directive, adopted in May 1992, calls for the establishment of Special Areas of Conservation (SACs), which, together with the Special Protection Areas (SPAs) designated under the 1979 Directive on the Conservation of Wild Birds, form a coherent European ecological network – Natura 2000. On the pollution side, the Water Act of 1989 introduced the concept of Statutory Quality Objectives (SQOs) for all controlled water including estuaries, whilst the move towards a generally more precautionary pollution control regime should have significant effects, such as the requirement for mandatory secondary treatment for estuarine sewage discharges above a certain size. Further, most major developments affecting the estuarine environment are now subject to environmental impact assessments (EIAs); if these are adequately conducted and implemented, protection of estuaries should be substantially advanced.

While most of these problems can be resolved at the local or regional levels, there is one looming issue which is global. For some years there has been concern about the increasing concentration of greenhouse gases, and whilst the position is as yet by no means clear, there are widely accepted forecasts of global warming and an associated sea level rise of perhaps 60 cm by the year 2100. If these predictions are accurate there will be a range of first-order effects in estuaries, including:

- increased frequency of flooding;
- rearrangement of unconsolidated sediment;

- changes in wave climates;
- accelerated dune and beach erosion;
- upstream intrusion of saltwater wedges;
- land-based retreat of the fresh/brackish water boundary.

To these may be added such second-order impacts as changes in bottom profiles, in sediment and nutrient fluxes and in primary production.

International assessment of the greenhouse issue has been taken up by several bodies. In particular, the WMO/UNEP Intergovernmental Panel on Climate Change (IPCC) has been examining present knowledge to produce a synthesis of scientific information and one of its most recent reports examines possible impacts (Houghton *et al.*, 1996). Many countries are taking the threat of climate change seriously, as shown by the 128 signatories of the UN Framework Convention on Climate Change (FCCC) established at the 1992 UN Conference on Environment and Development in Rio de Janeiro. Unfortunately, no effective action has so far been taken to reduce carbon dioxide emissions, and the most realistic response may be to examine the possibilities of adapting to the effects of climate change.

However, apart from these imponderables, we do appear to be approaching the new century with a useful strategy and the initiatives outlined above should go a long way towards ensuring that our estuaries can withstand the pressures likely to be imposed on them in the coming years. In general, we have got matters into perspective, recognizing that an integrated approach is required, and scientists, administrators, industrialists and politicians are beginning to work together towards the common end of ensuring that the estuarine resource is managed sustainably.

Given the extent and diversity of degradation recorded in estuaries around the world, it is not surprising that the topic of restoration is well highlighted. As discussed above, the first steps are of course to eliminate noxious inflows and to control damaging activities, but more positive and constructive approaches are possible. In this context, coastal engineering has much to offer (NRC, 1994). High level engineering works are of course much employed in the field of coastal protection, but are also relevant in attempts to reverse the loss and conversion of marine habitat, with (for example) the use of dredged material in habitat restoration work. In this approach it is important to note that a result which is entirely satisfactory from the engineering perspective may not leave the wildlife happy. Indeed, there may be conflicting views on what constitutes restoration, cosmetic development being totally out of keeping with the re-creation of a natural ecosystem. However, there are success stories from the United States, including accounts of saltmarsh restoration in Virginia and San Francisco and oyster bed restoration in Chesapeake Bay using dredged material (NRC, 1994).

Coming back finally to another success story and the subject of this book – the Thames – it is clear that this estuary will benefit, along with all the others, from the increasing awareness of environmental problems, from the enhanced monitoring techniques, and from the stricter and more relevant regulations introduced. One indicator of environmental improvement is a greater diversity of fauna and this we are now seeing. Otters (*Lutra lutra*), for example, disappeared from vast areas of the country with the onset of chemical pollution in the early 1960s, and are seen as an endangered species. But in 1996 an otter was observed exploring the Thames just 40 miles upstream from London, and it has been forecast that these animals will be seen on the river at Westminster early in the twenty-first century. The passage of salmon (*Salmo salar*) is another indicator of estuarine quality. The first salmon began returning to the Thames in 1974, but their establishment has been slow. However, in 1997 salmon were found in the river Medway, not having been recorded there for more than a hundred years, and the Salmon and Trout Association suggest that they may have spawned in the Thames. In the same year the Environmental Agency reported that other species (such as smelt, *Osmerus eperlanus*, and Dover sole, *Solea solea*; Chapter 7) had returned in healthy numbers to the estuary and in some cases further upstream, whilst in the summer of 1997 millions of juvenile sea bass (*Dicentrarchus labrax*) were detected under London Bridge. There is a long way to go, and continued surveillance will be essential, but at least the mechanism for protection and improvement is firmly in place.

Appendix A
Summary of data for selected sites along the Thames Estuary

(a) Physicochemical parameters

Site	National Grid Reference (TQ)	Symbol	Distance from Teddington weir (km)	Mean quarterly salinity (g/l)		Sediment description	Particle size analysis (%)						Depth (m)
				Max (quarter)	Min (quarter)		< 63 µm	63–212	212–600	600–1000	1–2 mm	> 2 mm	
Teddington	168 715	T	0.1	0.24 (4/91)[a]	0.08 (1/90)[a]	Shingle, stones on sand	1.77	5.99	10.37	2.52	3.21	76.14[b]	0.5
Kew	191 779	K	7.0	0.24 (3/90)	0.08 (1/90)	Stones, shingle, debris on muddy sand	9.29	12.44	36.24	8.17	5.14	28.72[b]	0.5
Hammersmith Bridge	230 780	HB	15.8	0.94 (3/90)	0.08 (1/90)	Sand and organic matter	7.35	16.45	74.74	0.66	0.32	0.48	Intertidal
Cadogan Pier	274 776	CP	23.2	2.38 (3/90)	0.08 (1/90)	Stones and debris on fine sand and mud	25.59	11.21	47.35	2.21	0.39	13.25[b]	0.5
South Bank Centre	308 803	SBC	28.3	3.32 (3/90)	0.08 (1/90)	Sand, organic matter and debris	2.27	9.89	85.20	1.80	0.61	0.23	Intertidal
London Bridge	327 805	LB	30.4	4.97 (3/90)	0.09 (1/90)	Large rocks on mud	24.78	20.26	22.59	2.38	5.83	24.16[b]	0.5
Greenwich	383 780	GW	38.3	6.82 (3/90)	0.14 (1/90)	Silt and mud	22.54	52.34	22.05	1.05	2.02	0.00	Intertidal
Woolwich i	427 793	WWi	45.1	12.20 (3/90)	0.97 (1/90)	Mud and stones	13.82	14.75	12.62	1.61	1.97	55.23	Intertidal
Woolwich s	429 794	WWs	45.6	"	"	Silt and mud	35.06	54.32	10.35	0.15	0.12	0.00	2.0
Beckton	456 815	BK	48.9	13.58 (3/90)	2.04 (1/90)	Fine mud	50.65	39.61	9.24	0.09	0.20	0.21	1.5
Crossness i	492 809	XNi	53.1	17.09 (3/90)	2.51 (1/90)	Coarse mud	34.76	48.32	16.25	0.25	0.42	0.00	Intertidal
Crossness s	494 809	XNs	53.3	"	"	Fine mud	40.05	50.76	8.46	0.11	0.20	0.42	2.0
Purfleet i	548 786	Pi	61.4	20.20 (4/89)	6.03 (1/90)	Fine mud	67.43	16.35	16.13	0.02	0.00	0.07	Intertidal
Purfleet s	580 761	Ps	65.3	"	"	Fine mud	41.19	36.54	22.12	0.15	0.00	0.00	2.0
West Thurrock i	592 770	WTi	66.6	24.57 (3/90)	9.88 (1/90)	Coarse mud	44.47	45.03	9.21	0.30	0.45	0.54	Intertidal
West Thurrock s	593 770	WTs	66.7	"	"	Coarse mud	36.10	33.99	12.07	0.27	2.03	4.54	2.0
Gravesend i	648 745	GVi	73.1	26.81 (3/90)	12.61 (1/90)	Sandy mud	22.54	51.21	26.19	0.06	0.00	0.00	Intertidal
Gravesend s	649 746	GVs	73.2	"	"	Coarse mud	38.36	52.54	8.57	0.53	0.00	0.00	3.0
Mucking	707 808	MK	83.0	31.41 (4/89)	20.73 (1/90)	Sandy mud	29.49	46.50	23.57	0.21	0.23	0.00	2.5
Blythe Sands	757 805	BS	86.9	"	"	Mud and shells	23.80	65.26	8.72	0.45	0.60	1.17	2.0
Canvey Beach	800 824	CB	91.7	32.95 (4/89)	25.96 (1/90)	Sandy mud	35.84	54.58	9.54	0.01	0.01	0.02	Intertidal
Chapman Buoy	809 813	CHB	93.1	"	"	Mud, stones, shells, sodden wood and clay	20.70	18.96	13.63	2.08	2.63	42.00	> 20.0
Allhallows	838 792	AH	95.1	"	"	Mud and shells	37.82	35.15	20.51	0.57	0.94	5.01	Intertidal
Grain Flats	877 795	GF	99.6	Generally > 30	"	Fine sand	20.79	47.03	31.14	0.02	0.02	0.00	3.0
Southend i	888 844	SEi	100.1	Generally > 30		Muddy sand and shells	14.44	65.82	12.92	0.78	0.51	5.53	Intertidal
Southend s	901 828	SEs	102.4	Generally > 30		Sand, stones and shell	5.83	29.42	52.34	0.39	0.48	11.54	15.0
Shoeburyness East	949 850	SNE	105.8	Generally > 30		Sand and shells	7.81	31.48	57.06	0.56	0.51	2.58	Intertidal
Sea Reach No. 2 Buoy	955 810	SR2	108.2	Generally > 30		Sand	5.71	50.50	43.28	0.29	0.07	0.15	12.0

[a] = Richmond

(b) Biological parameters

Site	Symbol	Sampling method and size	Macrofauna: Mean spp. No ± SD	Macrofauna: Mean abundance ± SD	Macrofauna: Mean biomass ± SD	Macrofauna: Abundance dominant(s)	Macrofauna: Biomass dominant	Meiofauna: Mean No. spp. ± SD	Meiofauna: Mean abundance ± SD	Meiofauna: Abundance dominant
Teddington	T	Kick 3 min	20.3 ± 6.6	1024 ± 500	10.3 ± 6.7	*G. zaddachi*	*G. zaddachi*	18.0 ± 7.4	7560 ± 6042	*Tobrilus gracilis*
Kew	K	Kick 3 min	15.1 ± 3.1	3096 ± 2000	23.2 ± 15.5	*G. zaddachi*	*G. zaddachi*	17.0 ± 3.4	3159 ± 755	*Tobrilus gracilis*
Hammersmith Bridge	HB	Inter 0.4 m^2	8.7 ± 2.9	280 ± 238	1.3 ± 1.2	*L. hoffmeisteri*	*G. zaddachi*	14.8 ± 4.6	387 ± 230	*Daptonema setosa*
Cadogan Pier	CP	Kick 3 min	12.0 ± 1.9	2298 ± 1081	24.2 ± 12.2	*P barbatus* and *G. zaddachi*	*G. zaddachi*	17.8 ± 3.9	2515 ± 247	*Daptonema setosa*
South Bank Centre	SBC	Inter 0.4 m^2	4.5 ± 1.5	71 ± 118	0.6 ± 0.8	*G. zaddachi*	*G. zaddachi*	8.5 ± 4.1	253 ± 108	*Daptonema setosa*
London Bridge	LB	Kick 3 min	6.4 ± 3.2	678 ± 649	11.0 ± 9.4	*G. zaddachi*	*G. zaddachi*	16.3 ± 3.9	3674 ± 2600	*Daptonema normandica*
Greenwich	GW	Inter 0.4 m^2	5.8 ± 2.5	1330 ± 1722	6.2 ± 13.3	*L. hoffmeisteri*	*L. hoffmeisteri*	19.3 ± 4.0	2775 ± 2455	None
Woolwich i	WWi	Inter 0.4 m^2	6.3 ± 2.5	403 ± 350	1.2 ± 0.8	*L. hoffmeisteri*	*L. hoffmeisteri*	12.3 ± 8.1	3116 ± 2850	*Daptonema setosa*
Woolwich s	WWs	Day grab 0.4 m^2	4.9 ± 2.3	192 ± 161	1.0 ± 0.9	*M. rubroniveus*	None	7.0 ± 4.3	419 ± 407	*Sabatieria punctata*
Beckton	BK	Day grab 0.4 m^2	2.2 ± 2.0	160 ± 393	0.7 ± 1.2	*L. hoffmeisteri*	*L. hoffmeisteri*	4.5 ± 2.5	128 ± 102	None
Crossness i	XNi	Inter 0.4 m^2	4.9 ± 1.6	10243 ± 8200	28.5 ± 20.5	*T. costatus*	*T. costatus*	12,8 ± 4.3	18198 ± 25198	*Daptonema setosa*
Crossness s	XNs	Day grab 0.4 m^2	6.2 ± 2.6	404 ± 388	1.8 ± 1.9	*M. rubroniveus*	*M. rubroniveus*	6.3 ± 4.0	329 ± 197	None
Purfleet i	Pi	Inter 0.4 m^2	5.7 ± 2.7	138 ± 122	1.2 ± 2.2	*M. rubroniveus*	*M. rubroniveus*	14.5 ± 3.0	3712 ± 3620	*Sabatieria punctata*
Purfleet s	Ps	Day grab 0.4 m^2	6.3 ± 3.3	1163 ± 1924	7.2 ± 11.6	*T. benedeni*	*C. volutator*	5.5 ± 2.7	2426 ± 3554	*Sabatieria punctata*
West Thurrock i	WTi	Inter 0.4 m^2	6.0 ± 1.5	6838 ± 3960	356.3 ± 165.6	*N. diversicolor*	*N. diversicolor*	19.3 ± 2.3	8160 ± 5772	*Sabatieria punctata*
West Thurrock s	WTs	Day grab 0.4 m^2	5.8 ± 2.4	2295 ± 6789	15.7 ± 42.9	*C. volutator*	*C. volutator*	7.8 ± 5.8	942 ± 808	*Sabatieria punctata*
Gravesend i	GVi	Inter 0.4 m^2	9.5 ± 1.5	5928 ± 3600	35.6 ± 39.9	*T. benedeni*	*T. benedeni*	22.0 ± 7.8	10097 ± 4357	*Daptonema* sp.
Gravesend s	GVs	Day grab 0.4 m^2	5.9 ± 3.2	590 ± 1553	5.5 ± 14.8	None	*C. volutator*	14.5 ± 3.6	1198 ± 1156	*Sabatieria punctata*
Mucking	MK	Day grab 0.4 m^2	8.5 ± 2.3	2630 ± 2728	11.9 ± 10.5	*Caulleriella*	*N. hombergi*	13.0 ± 2.5	11251 ±1136	*Sabatieria punctata*
Blythe Sands	BS	Day grab 0.4 m^2	6.3 ± 2.1	154 ± 115	3.0 ± 2.2	*N. hombergi*	*N. hombergi*	20.0 ± 4.7	4798 ± 6425	None
Canvey Beach	CB	Inter 0.4 m^2	14.5 ± 4.4	535 ± 174	25.6 ± 23.6	*N. hombergi*	*N. hombergi*	37.0 ± 3.7	39358 ± 16042	*Sabatieria punctata*
Chapman Buoy	CHB	Day grab 0.4 m^2	33.3 ± 9.2	1012 ± 375	248.3 ± 270.7	*S. troglodytes*	*S. troglodytes*	37.8 ± 12.1	8421 ± 5475	None
Allhallows	AH	Inter 0.4 m^2	16.9 ± 2.3	5994 ± 2285	74.3 ± 30.5	*T. benedeni*	*M. balthica*	32.8 ± 7.3	48385 ± 11105	*Ptycholaimellus ponticus*
Grain Flats	GF	Day grab 0.4 m^2	8.5 ± 2.3	314 ± 300	9.6 ± 6.9	*N. hombergi*	*N. hombergi*	21.5 ± 6.6	9539 ± 6518	*Sabatieria punctata*
Southend i	SEi	Inter 0.4 m^2	17.2 ± 3.0	2610 ± 2135	39.7 ± 21.8	*Caulleriella*	*C. edule*	36.3 ± 3.5	108357 ± 26726	None
Southend s	SEs	Day grab 0.4 m^2	20.5 ± 8.5	350 ± 143	10.6 ± 4.1	None	*N. hombergi*	28.8 ± 4.9	5536 ± 2990	*Richtersia inaequalis*
Shoeburyness East	SNE	Inter 0.4 m^2	16.5 ± 2.2	5111 ± 5750	54.7 ± 7.8	*H. ulvae*	*C. edule*	40.0 ± 5.3	77560 ± 31638	*Calomicrolaimus honestus*
Sea Reach No. 2 Buoy	SR2	Day grab 0.4 m^2	18.3 ± 4.9	447 ± 281	10.7 ± 7.0	*A minuta*	*N. cirrosa*	24.8 ± 9.7	7744 ± 1258	*Richtersia inaequalis*

Macrofauna abundance = I/m^2 (inter and Day grab), I/3 min (kick); biomass = gWW/m^2 or /3 min.
Meiofaunal abundance = I/l .
Intertidal macrofauna samples obtained using a 0.02 m^2 hand-operated Gully grab, five adjacent grabs giving one 0.1 m^2 replicate. For both intertidal and subtidal, four replicates were taken. Kick samples comprised 3 × 1 min kick.
Full Latin names: *Aricidia minuta*, *Caulleriella* sp., *Cerastoderma edule*, *Corophium volutator*, *Gammarus zaddachi*, *Hydrobia ulvae*, *Limnodrilus hoffmeisteri*, *Macoma balthica*, *Monopylephorus rubroniveus*, *Nephtys cirrosa*, *Nephtys hombergi*, *Nereis diversicolor*, *Psammoryctides barbatus*, *Sagartia troglodytes*, *Tubifex costatus*, *Tubificoides benedeni*.

Appendix B
Thames estuary species list

TEDDINGTON WEIR–SEA REACH NO. 2 BUOY

If there are any additions to this list, please contact Dr Martin Attrill at the address on the title page, giving species name, location and habitat of the species identified. Any further notes of interest would always be welcome.

ALGAE

Chlorophyceae

Tetrasporales	Tetrasporaceae	*Chaetopeltis orbicularis* Berthelot
Chlorococcales	Chlorococcaceae	*Chlorococcum infusionum* (Schrank) Meneghini
	Oocystaceae	*Chlorella* species
	Dictyosphaeriaceae	*Sphaerobotrys fluviatilis* Butcher
Oedogoniales	Oedogoniaceae	*Oedogonium* species
Chaetophorales	Chaetophoraceae	*Protoderma frequens* (Butcher) Printz
		Protoderma viride Kuetzing
		Stigeoclonium farctum Berthelot
		Stigeoclonium helvetica Vischer
		Stigeoclonium tenue Kuetzing
		Uronema confervicolum Lagerheim
Ulotrichales	Ulotrichaceae	*Eugomontia sacculata* Kornmann
		Hormidium flaccidum A. Braun
		Ulothrix flacca (Dillwyn) Thuret
		Ulothrix tenerrima Kuetzing
		Ulothrix tenuissima Kuetzing
		Ulothrix zonata (Weber & Mohr) Kuetzing
Ctenocladales	Ctenocladaceae	*Desmococcus olivaceus* (Acharius) Laundon

		Pseudendoclonium basiliense Vischer var. *brandii* Vischer
		Pseudendoclonium laxum D. John & L. Johnson
		Pseudendoclonium prostratum Tupa var. *pseudoprostratum* D. John & L. Johnson
		var. *radiatum* D. John & L. Johnson
Microthamniales	Microthamniaceae	*Microthamnium kuetzingianum* Naegeli
Klebsormidiales	Klebsormidiaceae	*Stichococcus bacillaris* Naegeli
Ulvales	Monostromataceae	*Blidingia marginata* (J. Agardh) P. Dangeard
		Blidingia minima (Kuetzing) Kylin
		Monostroma grevillei (Diliwyn) Wittrock
	Ulvaceae	*Enteromorpha flexuosa* (Roth) J. Agardh
		Enteromorpha intestinalis (L.) Link
		Enteromorpha prolifera (O.F. Mueller) J. Agardh
		Enteromorpha torta (Mertens) Reinbold
		Ulva lactuca L.
		Ulvaria oxysperma (Kuetzing)
		var. oxysperma
		f. *wittrockii* (Bornet) Bliding
Acrosiphoniales	Codiolaceae	*Urospora penicilliformis* (Roth) Areschoug
Cladophorales	Cladophoraceae	*Chaetomorpha capillaris* (Kuetzing) Boergesen
		Cladophora glomerata L.
		Cladophora fracta (Vahi) Kuetzing
		Cladophora sericea (Hudson) Kuetzing
		Rhizoclonium hieroglyphicum Kuetzing
		Rhizoclonium tortuosum (Dillwyn) Kuetzing
Bryopsidales	Bryopsidaceae	*Bryopsis plumosa* (Hudson) C. Agardh
Zygnematales	Zygnemataceae	*Spirogyra* species

Tribophyceae (Xanthophyceae)

Tribonematales	Tribonemataceae	*Tribonema bombycinum* (C. Agardh) Derbes & Solier
Vaucheriales	Vaucheriaceae	*Vaucheria bursata* (O.F. Mueller) C. Agardh
		Vaucheria canalicularis (L.) Christensen
		Vaucheria compacta (Collins) W.R. Taylor

Class Phaeophyceae

Ectocarpales	Ectocarpaceae	*Ectocarpus siliculosus* (Dillwyn) Lyngbye
		Ectocarpus sp.
		Pilayella littoralis (L.) Kjellman
	Scytosiphonaceae	*Petalonia fascia* (O.F. Mueller) Kuntze
		Stragularia clavata (Harvey) Hamel
	Elachistaceae	*Elachista fucicola* (Velley) Areschoug
Fucales	Fucaceae	*Ascophyllum nodosum* (L.) Le Jolis
		Fucus spiralis L.
		Fucus vesiculosus L.

Rhodophyceae

Bangiophycideae

Porphyridiales	Porphyridiaceae	*Porphyridium purpureum* (Bory) Drew & Ross
Bangiales	Bangiaceae	*Bangia atropurpurea* (Roth) C. Agardh
		Porphyra purpurea (Roth) C. Agardh
	Erythropeltidaceae	*Erythrotrichia carnea* (Dillwyn) C. Agardh
Nemaliales	Acrochaetiaceae	*Audouinella daviesii* (Dillwyn) Woelkerling
		Audouinella floridula (Dillwyn) Woelkerling
		Audouinella purpurea (Lightfoot) Woelkerling
		Audouinella sp.

	Batrachospermaceae	*'Chantransia*-stage'
	Gelidiaceae	*Gelidium pusillum* (Stackhouse) Le Jolis
Cryptonemiales	Dumontiaceae	*Dumontia contorta* (S.G. Gmelin) Ruprecht
Gigartinales	Gigartinaceae	*Chondrus crispus* Stackhouse
Ceramiales	Ceramiaceae	*Callithamnion hookeri* (Dillwyn) S.F. Gray
		Callithomnion roseum Harvey
		Ceramium deslongchampsii Chauvin
	Rhodomelaceae	*Polysiphonia nigrescens* (Hudson) Greville
		Polysiphonia urceolata (Dillwyn) Greville

PHYLUM PROTOZOA

Sarcodina

Foraminiferida	Elphidiidae	*Elphidium* species
	Hormosinidae	*Rheophax scottii* Chaster
	Nodosariidae	*Lagena* species
	Bolivinitidae	*Braziliana variabilis* (Williamson)
	Rotaliidae	*Ammonia beccarii* (L.) var. *tepida* Cushman
		Ammonia beccarii (L.) var. ?*batavus*
	Discorbidae	*Rosalina* species
Arcellinida	Centropyxidae	*Centropyxis aculeatus* (Ehrenberg)
Gromida	Euglyphidae	*Euglypha acanthophthora* (Ehrenberg)

Rhizopoda

Polythalamia		*Polystomella striato-punctata* (F. & M.)
		Miliolina sp
		Biloculina sp.
		Nonionina sp
		Planorbulina mediterranensis (d'Orb.)
		Lagena striata (d'Orb.)
Cystoflagellata		*Noctiluca miliaris* Surir
Dinoflagellata		*Ceratium furca* Ehbg.

Ceratium fusus Ehbg.
Ceratium macroceros Ehbg.
Ceratium tripos Nitzsch.
Peridinium depressum Bail.

Ciliata

Tintinnopsis beroidea Stein
Tintinnopsis campanula (Ehr.) Dad.
Tintinnopsis major Meunier
Tintinnopsis parva Merkle
Tintinnopsis parvula Jorgensen
Tintinnopsis turbolum (Meunier)
Ptychocylus urnula Jorgensen
Codonella lagenula (Clap. & Jorg.)
Stenosonella avellana (Meunier)
Stenosonella ventricosa (Clap. & Jorg.)
Stenosonella producta Meunier
Stenosonella steini Jorgensen

PHYLUM COELENTERATA

Subphylum Medusozoa

Class Hydrozoa

Athecata	Tubulariidae	*Tubularia indivisa* (L.)
		Cordylophora caspia Allman
	Hydridae	*Hydra* species
	Bougainvillidae	*Bougainvillea britannica* Forbes
	Hydractiniidae	*Podocoryne carnea* Sars
	Corynidae	*Sarsia turbulosa* (Sars)
Thecata	Campanularidae	*Clytia* species
		Laomedia flexuaso Hincks
		Obelia species
	Sertulariidae	*Sertularia argentea* L.
		Sertularia cupressino (L.)

Class Scyphozoa

Semaeostomae	Cyanidae	*Cyanea capillata* (L.)
	Pelagiidae	*Chrysaora hysoscella* (L.)
	Aureliidae	*Aurelia aurita* (L.)

Subphylum Anthozoa

Class Alcyonaria

Alcyonacea	Alcyonidae	*Alcyonium digitatum* (L.)

Class Zoantharia

Actinaria	Actiniidae	*Actinia equina* (L.)
		Tealia felina (L.)
	Metridiidae	*Metridium senile* (L.)
	Sagartiidae	*Sagartia troglodytes* (Price)
		?*Cereus pedunculatus* (Pennant)

PHYLUM CTENOPHORA

Class Tentaculata

Cydippida	Pleurobrachiidae	*Pleurobrachia pileus* (O.F. Müller)

Class Nuda

Beroida	Beroidae	*Beroë cucumis* Fabricus

PHYLUM PLATYHELMINTHES

Class Turbellaria

Rhabdocoela		*Phaenocora* species
Tricladida	Planariidae	*Planaria torva* (Müller)
		Polycelis tenuis Ijima
	Dugesiidae	*Dugesia polychroa* (Schmidt)
	Dendrocoelidae	*Dendrocoelum lacteum* (Müller)

PHYLUM NEMERTEA

Class Anopla

Heteronemertea	Lineidae	*Lineus* species
Palaeonemertea	Cephalothricidae	*Cephalothrix rufifrons* (Johnston)

Class Enopla

Hoplonemertea	Tetrastemmatidae	*Tetrastemma* species

PHYLUM NEMATOMORPHA

Gordioidea		*Gordius* species

PHYLUM ENTOPROCTA

	Pedicellinidae	*Pedicellina* species

PHYLUM SIPUNCULA

	Golfingiidae	*Golfingia ?margaritaceum* (Sars)

PHYLUM ROTIFERA

Class Monogononta

Plioma	Lecanidae	*Lecane* species
	Brachionidae	*Brachionus calciflorus* Pallas
		Brachionus quadridentatus Hermann
		Keratella cochlearis (Gosse)
		Keratella quadrata (Muller)
		Synchaeto triophthalmus Lauterborn
		Synchaeta vorax Rousselet

PHYLUM KINORHYNCHA

Homalorhagida	Pycnophyidae	*Pychnopyes* species
Cyclorhagida	Echinoderidae	*Echinoderes dujardini* Clarapede

PHYLUM GASTROTRICHA

	Chaetonotidae	*Chaetonotus* species

PHYLUM TARDIGRADA

Arthrotardigrada	Batillipedidae	*Batillipes mirus* Richters
Echiniscoidea	Echiniscidae	*Echiniscus testudo* Doyere

Eutardigrada	Macrobiotidae	*Macrobiotus dispar* Murray
		Macrobiotus hufelandii Murray

PHYLUM NEMATODA

Class Adenophorea

Dorylaimida	Dorylaimidae	Dorylaimid species 1 (?*Dorylaimus*)
		Dorylaimid species 2
		Dorylaimid species 3
		Dorylaimus crassus De Man
		Labronema species
	Longidoridae	*Longidorus macrosoma* Hooper
	Mononchidae	*Iotonchus* species
		Mononchus aquaticus Coetzee
	Qudsianematidae	*Allodorylaimus* species
		Labronemo species (?*vulvapapillatum*)
		Eudorylaimus acuticauda (De Man)
Enoplida	Alaimidae	*Alaimus* species
	Anoplostomatidae	*Anoplostomo viviparum* (Bastian)
		Chaetonema riemanni Platt
	Anticominidae	*Anticoma acuminata* (Eberth)
	Cryptonchidae	*Cryptonchus* species
	Enchelidiidae	*Belbolla teisseiri* (Luc & De Coninck)
		Calyptronema maxweberi (De Man)
		Eurystomina species
	Enoplidae	*Enoplus brevis* Bastian
		Enoplus communis Bastian
	Ironidae	*Ironus ignavus* Bastian
		Syringolaimus species
	Leptosomatidae	*Pseudocella coecum* (Ssweljev)
	Oncholaimidae	*Adoncholaimus fuscus* (Bastian)
		Adoncholaimus thalassophygas (De Man)
		Metoncholaimus scanicus (Allgen)
		Oncholaimellus calvadosicus (De Man)
		Oncholaimid species 1
		Oncholaimid species 2 (?*Viscosio*)
		Oncholaimus brachycercus De Man

		Oncholaimus campylocercoides De Coninck & Stekhoven
		Oncholaimus oxyuris Ditlesven
		Oncholaimus skawensis Ditlesven
		Viscosia cobbi Filipjev
		Viscosia elegans (Kreis)
		Viscosia glabra (Bastian)
		Viscosia species
		Viscosia viscosa (Bastian)
	Oxystominidae	*Halalaimus capitulatus* Boucher
		Halalaimus gracilis De Man
		Halalaimus isaitshikovi (Filipjev)
		Halalaimus longicaudatus (Filipjev)
		Nemanema cylindraticaudatum (De Man)
		Oxystomina asetosa (Southern)
		Oxystomina elongata (Butschli)
		Thalassoalaimus tardus De Man
	Pandolaimidae	*Pandolaimus* species
	Prismatolaimidae	*Prismatolaimus species (?verrucosus)*
		Prismatolaimus stenurus sensu Cobb
	Thoracostomopsidae	*Enoploides brunettii* Gerlach
		Enoplolaimus vulgaris De Man
		Mesacanthion diplechma (Southern)
	Tripyloididae	*Bathylaimus capacosus* Hopper
		Tobrilus gracilis (Bastian)
		Tobrilus species
		Tripyla affinis De Man
		Tripyloides gracilis (Ditlesven)
		Tripyloides marinus (Butschli)
		Tripyloides species
Trefusiida	Trefusiidae	*Halanonchus* species
		Rhabdocoma riemanni Jayasree & Warwick
		Trefusia longicaudata De Man
		Trefusia zostericola Allgen
		Trefusiid species
Araeolaimida	Plectidae	Plectid species (?*Paraplectonema*)
		Plectus granulosus Bastian
Chromadorida	Aegialoalaimidae	Aegialoalalimid species 1 (?*Aegialoalaimus*)
		Aegialoaolaimid species 2

	Southernia zosterae Allgen
Ceramonematidae	*Dasynemoides* species (?*albaensis*)
	Ceramonematid species
Chromadoridae	*Atrochromadora microlaima* (De Man)
	Chromadora macrolaima De Man
	Chromadorella species
	Chromadorid species
	Chromadorina bioculata (Schultze)
	Chromadorita leuckarti (De Man)
	Chromadorita nana Lorenzen
	Chromadorita species
	Chromadorita tentabunda (De Man)
	Dichromadora cephalata (Steiner)
	Dichromadora cucullata Lorenzen
	Dichromadora geophila (De Man)
	Dichromadora species
	Euchromadora vulgaris Bastian
	Hypodontolaimus balticus (Schneider)
	Hypodontolaimus inaequalis (Bastian)
	Hypodontolaimus species
	Innocuonema species
	Neochromadora poecilosoma (De Man)
	Neochromadora species
	Neochromadora tricophora (Steiner)
	Prochromadora species (?*orleji*)
	Prochromodorella attenuata (Gerlach)
	Prochromodorella ditlevseni (De Man)
	Prochromodorella septempapillata Platt
	Ptycholaimellus ponticus (Filipjev)
	Punctodora species
	Spilophorella candida Gerlach
	Spilophorella paradoxa (De Man)
	Spilophorella species
Comesomatidae	*Cervonema* species
	Comesomatid species
	Sabatieria breviseta Stekhoven
	Sabatieria celtica Southern
	Sabatieria longisetosa (Kreis)

	Sabatieria ornata (Ditlesven)
	Sabatieria praedatrix De Man
	Sabatieria pulchra (Schneider)
	Sabatieria punctata (Kreis)
	Setosabatieria species (?*hilarula*)
Cyatholaimidae	Cyatholaimid species 1 (?*Cyatholaimus*)
	Cyatholaimid species 2
	Cyatholaimus species (?*gracilis*)
	Marylynnia complexa (Warwick)
	Marylynnia species
	Paracanthonchus caecus Micoletzky
	Paracanthonchus heterodontus (Schulz)
	Paracanthonchus longus Allgen
	Paracanthonchus species
	Paracyatholaimus intermedius De Man
	Paracyatholaimus species (?*inglisi*)
	Paralongicyatholaimus species
	Pomponema species
	Praecanthonchus ophelioe (Warwick)
	Praecanthonchus species
Desmodoridae	*Chromaspirina* species
	Desmodora communis (Butschli)
	Desmodora species
	Desmodorid species 1 (?*Chromaspirina*)
	Desmodorid species 2
	Metachromadora remanei Gerlach
	Metachromadora scotlandica Warwick & Platt
	Metachromadora species
	Metachromadora suecica (Allgen)
	Metachromadora vivipara (De Man)
	Molgolaimus cuanensis (Platt)
	Onyx perfectus Cobb
	Pseudonchus species
	Sigmophoranema rufum (Cobb)
	Spirinia parasitifera (Bastian)
Leptolaimidae	*Antomicron elegans* (De Man)
	Camacolaimus barbatus Warwick
	Camacolaimus longicauda De Man

		Camacolaimus tardus De Man
		Chronogaster species
		Deontolaimus species
		Leptolaimid species
		Leptolaimoides species
		Leptolaimus ampullaceus Warwick
		Leptolaimus elegans (Stekhoven & De Coninck)
		Leptolaimus papilliger De Man
		Leptolaimus species 2 (?*limicolus*)
		Leptolaimus species 3
		Onchium conicaudatus (Allgen)
		Stephanolaimus jayassrei Plan
		Stephanolaimus species (?*spartinae*)
	Microlaimidae	*Aponema torosa* (Lorenzen)
		Calomicrolaimus honestus (De Man)
		Calomicrolaimus parahonestus (Gerlach)
		Microlaimus conothelis (Lorenzen)
		Microlaimus globiceps De Man
		Microlaimus marinus (Schulz)
		Microlaimus robustidens Stekhoven & De Coninck
	Monoposthiidae	*Monoposthia castata* (Bastian)
		Monoposthia mirabilis Schulz
	Selachinematidae	*Gammanema rapax* (Ssweljev)
		Halichoanolaimus robustus (Bastian)
		Richtersia inaequalis Riemann
Desmoscolecida	Desmoscolecidae	*Desmoscolex falcatus* Lorenzen
		Pareudesmoscolex species
		Quadricoma species
Monhysterida	Axanolaimidae	*Ascolaimus elongatus* (Butschli)
		Axonolaimus paraspinosus Stekhoven & Adam
		Axonolaimus species
		Odontophora setosa (Allgen)
		Odontophora villoti (Luc & De Coninck)
	Coninckiidae	*Coninckia* species
	Diplopeltidae	*Campylaimus* species (?*inaequalis*)
		Diplopeltid species (?*Diplopeltula*)
	Linhomoeidae	*Desmolimus* species
		Desmolaimus zeelandicus De Man

		Eleutherolaimus species
		Linhomoeid species 1 (?*Terschellingia*)
		Linhomoeid species 2
		Linhomoeid species 3
		Linhomoeus species 1
		Linhomoeus species 2
		Terschellingia communis De Man
		Terschellingia longicaudata De Man
		Terschellingia species
	Monhysteridae	*Diplolaimello ocellata* Chitwood
		Monhystera filicaudata Allgen
		Monhystera disjuncta Bastian
		Monhystera species
		Monhystera stagnalis Bastian
		Monhystera vulgaris De Man
		Monhysterid species 1
		Monhysterid species 2
	Sphaerolaimidae	*Parasphaeralaimus* species (??*paradoxa*)
		Sphaerolaimus balticus Schneider
		Sphaerolaimus gracilis De Man
	Xyalidae	*Daptonema normandica* (De Man)
		Daptonema setosa (Butschli)
		Daptonema tenuispiculum (Ditlesven)
		Daptonema species 1
		Daptonema species 2
		Linhystera species
		Paramonhystera species
		Theristus acer Bastian
		Theristus species 1
		Theristus species 2
		Theristus species 3
		Xyalid species 1
		Xyalid species 2

Class secernentea

Ascaridida	Anisakidae	*Contracaecum aduncum* (Rudolphi)
		Terranova species
	Cucullanidae	*Cucullanus heterochrous*
Rhabditida	Cephalobidae	*Acrobeles* species
		Cephalobus species
	Diplogasteridae	*Butlerius micans* Pillai & Taylor

		Diplogaster species
		Diplogasterid species
		Paroigolaimella bernensis (Steiner)
	Panagrolaimidae	*Panagrolaimus* species
	Rhabditidae	*Diploscapter coronatus* (Cobb)
		Mononchoides species
		Mononchoides striatus (Butschli)
		Panagrellus species
		Rhabditid species 1
		Rhabditid species 2
		Rhabditis species
Spirurida	Cystidicolidae	*Cystidicola* species
Tylenchida	Criconematidae	*Criconema* species
		Criconemoides amorphus De Grisse
		Macroposthonia involuta Loof
	Dolichodoridae	*Macrotrophurus arbusticola* Loof
	Heteroderidae	Heteroderid species (?*Globodera*)
	Hoplolaimidae	*Hirschmanniella* species
	Tylenchidae	*Halenchus fucicola* (De Man & Barton)
		Tylenchid species
		Tylenchus elegans De Man

PHYLUM ANNELIDA

Class Polychaeta

Orbiniida	Orbiniidae	*Scoloplos armiger* (Müller)
	Paraonidae	*Aricidea minuta* Southward
Spionida	Spionidae	*Malacoceros fuliginosus* (Claparède)
		Malacoceros tetracerus (Schmarda)
		Polydora ciliata (Johnston)
		Polydora flava Claparède
		Pygospio elegans (Claparède)
		Spio filicornis (Müller)
		Spiophanes bombyx (Claparède)
		Streblospio shrubsolii (Buchanan)
	Magelonidae	*Magelona mirabilis* (Johnston)
	Cirratulidae	*Caulleriella caput-esocis* (St-Joseph)
		Caulleriella zetlandica (McIntosh)
		Chaetozone setosa Malmgren
		Cirratulus filiformis Keferstein

		Tharyx marioni (St-Joseph)
Capitellida	Capitellidae	*Capitella capitata* (Fabricus)
		Capitomastus giardi (Mesnil)
		Notomastus latericeus M. Sars
	Arenicolidae	*Arenicola marina* (L.)
Opheliida	Opheliidae	*Ophelia ?neglecta* (Schneider)
Phyllodocida	Phyllodocidae	*Anaitides groenlandica* (Oersted)
		Anaitides maculata (L.)
		Anaitides mucosa (Oersted)
		Anaitidess rosea
		Eteone flava Fabricus
		Eteone longa (Fabricus)
		Eulalia bilineata Johnston
		Eumida sanguinea (Oersted)
		Mysta picta (Quatrefages)
		Notophyllum foliosum (Sars)
	Aphroditidae	*Aphrodite aculeata* L.
	Polynoidae	*Gattyana cirrhosa* (Pallas)
		Harmathoë impar Johnston
		Harmathoë extenuata (Grube)
		Lepidonotus squamatus (L.)
		Malmgrenia castanea McIntosh
	Sigalionidae	*Pholoë minuta* Fabricus
		Pholoë synophthalmica Claparède
		Sthenelais boa (Johnston)
	Syllidae	*Autolytus prolifera* (Müller)
		Eusyllis blomstrandi Malmgren
		Syllis gracilis Grube
		Syllis species
	Nereidae	*Nereis diversicolor* Müller
		Nereis longissima (Johnston)
		Nereis succinea (Frey & Leuckart)
		Nereis virens (Sars)
		Websterinereis glauca (Claparède)
	Glyceridae	*Glycera tridactyla* Schmarda
	Goniadidae	*Goniada maculata* Oersted
	Nephthyidae	*Nephthys caeca* Fabricus
		Nephthys cirrosa Ehlers
		Nephthys hombergi Savigny
		Nephthys longosetosa Oersted
Flabelligerida	Flabelligeridae	*Flabelligera affinis* Sars
Terebellida	Sabellariidae	*Sabellaria spinulosa* Leuckart
	Pectinariidae	*Lagis koreni* (Malmgren)
	Ampharetidae	*Ampharete acutifrons* Grube
	Terebellidae	*Lanice concheliga* (Pallas)

		Neoamphitrite figulus (Dalyell)
		Polycirrus aurantiacus Grube
		Thelpus cincinnatus (Fabricus)
		Thelpus ?setosus (Quatrefages)
Sabellida	Sabellidae	*Manayunkia aestuarina* (Bourne)
		Sabella pavonina Savigny
	Seropulidae	*Mercierella enigmatica* Fauvel
		Pomatocerss triqueter (L.)

Class Clitellata

Subclass Oligochaeta

Plesiopora	Naididae	*Dero digitata* (Müller)
		Nais elinguis Müller
		Paranais litoralis (Müller)
		Stylaria lacustris (L.)
		Uncinais uncinata (Ørstedt)
	Enchytraeidae	*Enchytraeus* species
	Tubificidae	*Aulodrilus pluriseta* (Piguet)
		Branchiura sowerbyi Beddard
		Clitellio arenarius (Müller)
		Limnodrilus cervix Brinkhurst
		Limnodrilus claparedeianus Ratzel
		Limnodrilus hoffmeisteri Claparède
		Limnodrilus udekemianus Claparède
		Monopylephorus irroratus (Verrill)
		Monopylephorus rubroniveus (Levinsen)
		Potamothrix bavaricus (Oschmann)
		Potamothrix hammoniensis (Michaelsen)
		Potamothrix moldaviensis (Vejdovsky & Mràzek)
		Psammoryctides barbatus (Grube)
		Ryacodrilus coccineus (Vejdovsky)
		Tubifex costatus Claparède
		Tubifex ?ignotus (Štolc)
		Tubifex tubifex (Müller)
		Tubificoides benedeni (Udekem)
		Tubificoides psuedogaster (Dahl)
	Lumbriculidae	*Lumbriculus variegatus* (Müller)
		Stylodrilus heringianus Claparède
	Lumbricidae	*Eiseniella tetraedra* (Savigny)

Subclass Hirudinea

Rhynchobdellae	Piscicolidae	?*Oceanobdella blennii* (Knight-Jones)
		Piscicola geometra (L.)
	Glossiphoniidae	?*Batracobdella paludosa* (Carena)
		Glossiphonia complanata (L.)
		Glossiphonia ?heteroclita (L.)
		Helobdella stagnalis (L.)
		Theromyzon tessulatum (O.F. Müller)
Pharyngobdellae	Erpobdellidae	*Erpobdella octoculata* (L.)
		Erpobdella testacea (Savigny)

PHYLUM CRUSTACEA

Class Branchiopoda

Cladocera	Chydoridae	*Pleuroxus uniciatus* Baird
	Daphniidae	*Bosmina longirostris* O.F. Muller
		Daphnia longispina O.F. Muller
		Eurycercus lamellatus (Müller)
		Ilyocryptus sordidus (Lieven)
		Sarsiella zostericola Cushman
		Simocephalus vetulus (Müller)

Class Ostracoda

Candona species

Class Copepoda

Calanoida	Temoridae	*Acartia bifilosa* Giesbrecht
		Acartia clausi Giesbrecht
		Acartia discaudata Giesbrecht
		Acartia longiremis (Lilljeborg)
		Eurytemora affinis (Pope)
		Eurytemora americana Williams
		Eurytemora velox (Lilljeborg)
		Temora longicornis (O.F. Muller)
Cyclopoida	Cyclopidae	*Acanthocyclops viridus* (Jurine)
		Cyclopina littoralis Boeck
		Cyclops strenuus (Fischer)
		Cyclops vernalis (Fischer)

		Cyclops vicinus Uljanin
		Eucyclops agilis (Koch)
		Eucyclops serrulatus Lilljeborg
		Hemicyclops purpureus Boeck
		Macrocyclops albidus (Jurine)
		Paracyclops fimbriatus (Fischer)
		Ascomyzon asterocheres Boeck
		Corycaeus anglicus Lubbock
		Oncea venusta Phillippi
Harpacticoida	Ameiridae	*Leptomesochra macintoshi* (T & A Scott)
	Canthocamptidae	*Attheyella* species
		Bryocamptus species (?*echinatus*)
		Canthocamptus species
		Elaphoidella gracilis (Sars)
		Epactophanes richardi Mrazek
		Mesochra lilljeborgi Boeck
		Moraria species
	Canuellidae	*Canuella perplexa* T & A Scott
	Cletodidae	*Cleta* species
		Cletodes longicaudatus (Boeck)
		Cletodes species
		Cletodid species
		Enhydrosoma propinguum (Brady)
		Itunella species
	Diosaccidae	*Amphiascella* species
		Amphiascoides species
		Amphiascus angusticeps Gurney
		Amphiascus longicornis (Claus)
		Amphiascus tenellus Sars
		Amphiascus species 3
		Amphiascus varians (Norman & Scott)
		Bulbamphiascus species
		Diosaccid copepodites
		Paramphiascella species
		Pseudomesochra species (?*latifurea*)
		Schizopera clandestina (Klie)
		Stenhelia aemula (T Scott)
		Stenhelia giesbrechti T & A Scott
		Stenhelia palustris Brady
		Stenhelia species
		Typhlamphiascus species
	Ectinosomatidae	*Arenostella* species
		Ectinosoma curticone Boeck

		Ectinosoma melaniceps Boeck
		Ectinosoma sarsi Boeck
		Halectinosoma curticorne Boeck
		Halectinosoma herdmani T & A Scott
		Halophytophilus species
		Pseudobradya brevicornis (T Scott)
		Pseudobradya similis (T & A Scott)
	Harpacticidae	*Harpacticella* species
	Laophontidae	*Laophonte cornuta* Phillipi
		Laophonte denticornis T Scott
		Laophontodes species
		Onychocamptus species (?*bengalensis*)
	Longipediidae	*Longipedia coronata* Claus
		Longipedia species
	Tachidiidae	*Microarthridion littorale* Poppe
		Tachidiid species
		Tachidius brevicornis Lilljeborg
		Tachidius discipes Giesbrecht
	Thalestridae	*Idomene forficata* Philippi
		Phyllothalestris species
	Tisbidae	*Tisbe finmarchica* (Sars)
		Tisbe furcata (Baird)

Class Branchiura

		Argulus foliaceus Nettowich

Class Cirripedia

Thoracica	Balanidae	*Balanus improvisus* Darwin
		Elminius modestus Darwin
		Semibalanus balanoides (L.)
Rhizocephala	Sacculinidae	*Sacculina carcini* Thompson

Class Malacostraca

Mysidacea	Mysidae	*Gastrosaccus normani* G.O. Sars
		Gastrosaccus sanctus (van Beneden)
		Gastrosaccus spinifer (Goes)
		Mesopodopsis slabberi (Van Beneden)
		Neomysis integer (Leach)

		Praunus flexuosus (Müller)
		Schistomysis ornata (G.O. Sars)
		Schistomysis spiritus (Norman)
		Siriella armata (Milne-Edwards)
		Siriella clausi G.O. Sars
Cumacea	Bodotriidae	*Bodotria scorpioides* (Montagu)
		Cumopsis goodsiri (Van Beneden)
	Diastylidae	*Diastylis bradyi* Norman
		Diastylis neopolitana G.O. Sars
	Pseudocumidae	*Pseudocuma longicornis* (Bate)
Isopoda	Gnathiidae	*Gnathia maxillaris* (Montagu)
		Paragnathia formica (Hesse)
	Anthuridae	*Cyanthura carinata* (Kröyer)
	Sphaeromatidae	*Sphaeroma monodi* Bocquet
		Sphaeroma rugicauda Leach
	Limnoriidae	*Limnoria lignorum* (Rathke)
	Idoteidae	*Idotea balthica* (Pallas)
		Idotea chelipes (Pallas)
		Idotea granulosa Rathke
		Idotea linearis (L.)
	Janiridae	*Jaera albifrons* gp. Leach
	Asellidae	*Asellus aquaticus* (L.)
	Ligiidae	*Ligia oceanica* (L.)
Amphipoda	Acanthonotozomatidae	*Panoploea minuta* (Sars)
	Stenothoidae	*Stenothoe marina* (Bate)
	Talitridae	*Orchestia gammarellus* (Pallas)
		Orchestia mediterranea Costa
		Talitrus saltator (Montagu)
	Hyalidae	*Hyale nilssoni* (Rathke)
	Gammaridae	*Chaetogammarus marinus* (Leach)
		Chaetogammarus ?stoerensis (Reid)
		Eulimnogammarus obtusaus (Dahl)
		Gammarus duebeni Liljeborg
		Gammarus lacustris Sars
		Gammarus locusta (L.)
		Gammarus pulex (L.)
		Gammarus salinus Spooner
		Gammarus zaddachi Sexton
	Crangonyctidae	*Crangonyx pseudogracilis* Bousfield
	Melitidae	*Melita obtusata* (Montagu)
		Melita palmata (Montagu)
	Haustoriidae	*Bathyporeia elegans* Watkin
		Bathyporeia guilliamsoniana (Bate)
		Bathyporeia pelagica (Bate)

		Bathyporeia pilosa Linström
		Bathyporeia sarsi Watkin
		Urothoe poseidonis Reibisch
	Oedicerotidae	*Monoculodes carinatus* (Bate)
		Periculodes longimanus (Bate & Westwood)
		Pontocrates altamarinus (Bate & Westwood)
	Calliopiidae	*Calliopius laeviusculus* (Kröyer)
	Atylidae	*Atylus falcatus* Metzger
		Atylus guttatus (Costa)
		Atylus swammerdami (Milne-Edwards)
		Atylus vedlomensis (Bate & Westwood)
	Aoridae	*Aora typica* Kröyer
	Isaeidae	*Microprotopus longiramis* (Levereux)
		Microprotopus maculatus Norman
	Corophiidae	*Corophium arenarius* Crawford
		Corophium ascherusicum (Costa)
		Corophium bonelli Milne-Edwards
		Corophium curvispinum Sars
		Corophium insidiosum Crawford
		Corophium lacustre Vanhöffen
		Corophium sextonae Crawford
		Corophium volutator (Pallas)
	Caprellidae	*Caprella linearis* (L.)
Euphausiacea	Euphausiidae	*Meganyctiphanes norvegica* (M. Sars)
		Nyctiphanes couchi (Bell)
Decapoda	Palaemonidae	*Palaemon adspersus* (Rathke)
		Palaemon elegans Rathke
		Palaemon longirostris Milne-Edwards
		Palaemon serratus (Pennant)
		Palaemonetes varians (Leach)
	Hippolytidae	*Hyppolyte varians* Leach
		Thoralus cranchi (Leach)
	Processidae	*Processa canaliculata* Leach
		Processa nouveli holthuisi Al-Adhub & Williamson
	Pandalidae	*Pandalina brevirostris* (Rathke)
		Pandalus montagui Leach
	Crangonidae	*Crangon crangon* (L.)
		Philocheras trispinosus (Hailstone)

	Porcellanidae	*Pisidia longicornis* (L.)
	Paguridae	*Pagurus bernhardus* (L.)
		Pagurus pubescens Kröyer
	Nephropidae	*Homarus gammarus* (L.)
	Majidae	*Hyas arenarius* (L.)
		Macropodia longirostris (Fabricus)
		Macropodia rostrata (L.)
	Corystidae	*Corystes cassivelaunus* (Pennant)
	Atelecyclidae	*Atelecyclus rotundus* (Olivi)
	Cancridae	*Cancer pagurus* L.
	Portunidae	*Carcinus maenas* (L.)
		Liocarcinus depurator (L.)
		Liocarcinus holsatus (Fabricus)
		Liocarcinus puber (L.)
	Pinnotheridae	*Pinnotheres pisum* (L.)
	Grapsidae	*Eriocheir sinensis* Milne-Edwards

PHYLUM CHELICERATA

Class Pycnogonida

Pantopoda	Nymphonidae	*Nymphon rubrum* Hodge
	Ammotheidae	*Achelia echinata* (Hodge)
	Phoxichilidiidae	*Anoplodactylus petiolatus* (Kröyer)
	Pycnogonidae	*Pycnogonum littorale* (Strom)

Class Arachnida

Acari	Limnesidae	Limnesid species
	Mideopsidae	Mideopsid species
	Bdellidae	Bdellid species
	Oribatidae	Oribatid species
	Hydracarina	Hydracarine species
	Halacaridae	*Copidognathus dentatus* Viets
		Copidognathus rhodostigma (Gosse)
		Thalassarachna baltica (Lohmann)

PHYLUM UNIRAMIA

Subphylum Myriapoda

Class Chilopoda

Geophilomorpha		*Strigamia maritima* (Leach)

Subphylum Hexapoda

Class Apterygota

Collembola	Machilidae	*Petrobius* species

Class Pterygota

Ephemeroptera	Baetidae	*Baetis* species
		Cloeon dipterum (L.)
	Ephemeridae	*Ephemera danica* Müller
	Caenidae	*Caenis horaria* (L.)
		Caenis moesta Bengtsson
Odonata	Coenagriidae	*Ischnura elegans* (Linden)
Hemiptera	Aphelocheiridae	*Aphelocheirus aestivalis* (Fabricus)
	Corixidae	*Sigara dorsalis* (Leach)
		Sigara falleni (Fieber)
Coleoptera	Haliplidae	*Haliplus* species
	Dytiscidae	*Deronectes depressus* (Fabricus)
	Elminthidae	*Oulimnius tuberculatus* (Müller)
Trichoptera	Rhyacophilidae	*Agapetus* species
	Psychomyiidae	*Tinodes waeneri* (L.)
	Hydropsychidae	*Hydropsyche* species
	Hydroptilidae	*Hydroptila* species
	Leptoceridae	*Athripsodes cinereus* (Curtis)
		Athripsodes nigronervosus (Retzius)
		Mystacides azurea (L.)
		Mystacides longicornis (L.)
Diptera	Tipulidae	*Erioptera* species
	Psychodidae	Psychodidae species
	Chaoboridae	*Chaoborus* species
	Ceratopogonidae	Ceratopogonidae species
	Chironomidae	*Brillia longiforca* Kieffer
		Bryophaenocladius species
		Chironomus species
		Conchapelopia melanops Miegen
		Cricotopus bicinctus (Meigen)
		Cricotopus species
		Cryptochironomus species
		Demeijerea species
		Dicrotendipes species
		Glyptotendipes species
		Harnischia species
		Isocladius sylvestris (Fabricus)
		Limnophyes species

		Metriocnemus species
		Micropsectra atrofasciata Kieffer
		Micropsectra species
		Microtendipes species
		Nanocladius species
		Orthocladius species
		Parachironomus species
		Paratendipes species
		Paratrichocladius species
		Phaenopsectra species
		Polypedilum species 1
		Polypedilum species 2
		Procladius species
		Prodiamesa olivacea (Meigen)
		Psectrocladius species
		Pseudosmittia species
		Rheocricotopus species
		Rheopelopia species
		Rheotanytarsus species
		Smittia species
		Synorthocladius species
		Thalassosmittia species
		Thienemannimyia species
		Xenochironomus species
	Tabanidae	Tabanidae species

PHYLUM MOLLUSCA

Class Polyplacophora

Neoloricata	Lepidochitonidae	*Lepidochitona cinerea* (L.)
	Lepidopleuridae	*Lepidopleurus asellus* (Gmelin)

Class Gastropoda

Subclass Prosobranchia

Archeogastropoda	Patellidae	*Patella vulgata* L.
	Trochidae	*Gibbula cineraria* (L.)
		Gibbula umbilicalis (da Costa)
	Neritidae	*Theodoxus fluviatilis* (L.)
Mesogastropoda	Littorinidae	*Littorina littorea* (L.)
		Littorina mariae Sacchi & Rastelli
		Littorina obtusata (L.)

		Littorina rudis (Maton)
	Hydrobiidae	*Bithynia tentaculata* (L.)
		Hydrobia ulvae (Pennant)
		Potamopyrgus jenkinsi (Smith)
	Assimineidae	*Assiminea grayana* Fleming
	Calyptraeidae	*Crepidula fornicata* (L.)
	Viviparidae	*Viviparus viviparus* (L.)
	Valvatidae	*Valvata piscinalis* (Müller)
Neogastropoda	Muricidae	*Nucella lapillus* (L.)
		Urosalpinx cinerea (Say)
	Buccinidae	*Buccinum undatum* L.

Subclass Opisthobranchia

Bullomorpha	Retusidae	*Retusa obtusa* (Montagu)
Nudibranchia	Onchidorididae	*Acanthodoris pilosa* (Abilgaard)
		Onchidoris bilamellata (L.)
	Archidorididae	*Archidoris pseudoargus* (Rapp)
	Facelinidae	*Facelina auriculata* (Müller)
	Aeolidiidae	*Aeolidia papillosa* (L.)

Subclass Pulmonata

Basommatophora	Ellobiidae	*Phytia myosotis* (Draparnaud)
	Lymnaeidae	*Lymnaea auricularia* (L.)
		Lymnaea peregra (Müller)
		Lymnaea stagnalis (L.)
	Physidae	*Physa heterostropha* (Say)
		Physa fontinalis (L.)
	Planorbidae	*Planorbis albus* Müller
		Planorbis carinatus Müller
	Ancylidae	*Acroloxus lacustria* (L.)
		Ancylus fluviatilis Müller

Class Bivalvia

Subclass Palaeotaxodonta

Nuculoida	Nuculidae	*Nucula sulcata* Bronn
		Nucula turgida Leckenby & Marshall

Subclass Pteriomorphia

Mytiloida	Mytilidae	*Modioluna phaseolina* (Philippi)

		Modiolus modiolus (L.)
		Mytilus edulis (L.)

Subclass Palaeoheterodonta

Unionacea	Unionidae	*Anodonta complanata* Rossmäller
		Anodonta cygnaea (L.)

Subclass Heterodonta

Veneroida	Montacutidae	*Mysella bidentata* (Montagu)
	Cardiidae	*Cerastoderma edule* (L.)
	Veneridae	*Venerupis pullastra* (Montagu)
	Petricolidae	*Petricola pholadiformis* Lamark
	Mactridae	*Mactra stultorum* (L.)
	Tellinidae	*Fabulina fabula* (Gmelin)
		Macoma balthica (L.)
		Moerella pygmaea (Lovén)
	Scrobicularidae	*Abra alba* (Wood)
		Abra nitida (Müller)
		Abra tenuis (Montagu)
		Scrobicularia plana (da Costa)
	Solenidae	*Ensis ?arcuatus* (Jeffreys)
	Sphaeriidae	*Pisidium* species
		Sphaerium corneum (L.)
		Sphaerium lacustre (Müller)
	Dreissenidae	*Dreissena polymorpha* (Pallas)
Myoida	Myacidae	*Mya arenaria* L.
	Pholadidae	*Barnea candida* (L.)
	Teredinidae	*Teredo navalis* L.

Class Cephalopoda

Subclass Coleoidea

Sepioidea	Sepiidae	*Sepia officianalis* L.
	Sepiolidae	*Sepiola atlantica* d'Orbigny
Teuthoidea	Loliginidae	*Alloteuthis subulata* (Lamark)

PHYLUM BRYOZOA

Gymnolaemata	Menbraniporidae	*Conopeum seurati*
		Electra crustulenta
		Electra pilosa (L.)

		Farella repens
Ctenostomata	Alcyonidiidae	*Alcyonidium gelatinosum* (L.)
		Alcyonidium hirsutum (Fleming)
Cheilostomata	Flustridae	*Flustra foliacea* (L.)

PHYLUM CHAETOGNATHA

		Sagitta elegans Verril
		Sagitta setosa J. Muller

PHYLUM ECHINODERMATA

Subphylum Asterozoa

Class Stelleroidea

Subclass Asteroidea

Spinulosida	Solasteridae	*Crossaster papposus* (L.)
Forcipulatida	Asteriidae	*Asterias rubens* L.

Subclass Ophiuroidea

Ophiurae	Ophiolepidae	*Ophiura ophiura* (L.)
	Ophiotrichidae	*Ophiothrix fragilis* (Abildgaard)
	Amphiuridae	*Amphipholis squamata* (Delle Chiaje)

Subphylum Echinozoa

Class Echinoidea

Echinacea	Echinidae	*Psammechinus miliaris* (Gmelin)
Gnathostomata	Fibulariidae	*Echinocyamus pusillus* (O.F. Müller)
Atelostomata	Spatangidae	*Echinocardium cordatum* (Pennant)

PHYLUM CHORDATA

Subphylum Urochordata

Class Ascidiacea

Peurogona	Molgulidae	*Molgula manhattensis* (De Kay)

Subphylum Vertebrata

Superclass Agnatha

Class Cyclostomata

Hyperoartia	Petromyzonidae	*Lampetra fluviatilis* (L.)
		Lampetra planeri (Bloch)

Superclass Gnathostomata

Class Selachii

Pleurotremata	Scyliorhinidae	*Scyliorhinus caniculus* (L.)
	Triakudae	*Mustelus mustelus* (L.)
Hypotremata	Rajidae	*Raja clavata* L.
		Raja undulata Lacépède
	Dasyatidae	*Dasyatis pastinaca* (L.)

Class Osteichthyes

Isopondyli	Clupeidae	*Alosa alosa* (L.)
		Alosa fallax (Lacépède)
		Clupea harengus L.
		Engraulis encrasicolus (L.)
		Sardina pilchardus (Walbaum)
		Sprattus sprattus (L.)
	Thaymallidae	*Thymallus thymallus* (L.)
	Salmonidae	*Oncorhynchus mykiss* (Walbaum)
		Salmo salar L.
		Salmo trutta L.
	Osmeridae	*Osmerus eperlanus* (L.)
Haplomi	Esocidae	*Esox lucius* L.
Ostariophysi	Cyprinidae	*Abramis brama* (L.)
		Alburnus alburnus (L.)
		Barbus barbus (L.)
		Carassius auratus (L.)
		Carassius carassius (L.)
		Cyprinus carpio L.
		Gobio gobio (L.)
		Leuciscus cephalus (L.)
		Leuciscus leuciscus (L.)

		Phoxinus phoxinus (L.)
		Rutilus rutilus (L.)
		Scardinius erythrophthalmus (L.)
		Tinca tinca (L.)
	Cobitidae	*Noemacheilus barbatulus* (L.)
Apodes	Anguillidae	*Anguilla anguilla* (L.)
	Congridae	*Conger conger* (L.)
Synentognathi	Scomberesocidae	*Scomberesox saurus* (Walbaum)
	Belonidae	*Belone bellone* (L.)
Solenichthyes	Syngnathidae	*Hippocampus hippocampus* L.
		Hippocampus ramulosus Leach
		Nerophis lumbriciformis (Jenyns)
		Sygnatus acus L.
		Sygnatus rostellatus Nilsson
		Sygnatus typhle L.
Anacanthini	Gadidae	*Ciliata mustela* (L.)
		Ciliata septentrionalis (Collett)
		Gadus morhua L.
		Gaidropsarus mediterraneus (L.)
		Gaidropsarus vulgaris (Cloquet)
		Melanogrammus aeglefinus (L.)
		Merlangius merlangus (L.)
		Merluccius merluccius (L.)
		Micromesistius poutassou (Risso)
		Molva molva (L.)
		Pollachius pollachius (L.)
		Raniceps raninus (L.)
		Rhinonemus cimbrius (L.)
		Trisopterus luscus (L.)
		Trisopterus minuta (L.)
Zeomorphi	Zeidae	*Zeus faber* L.
Percomorphi	Serranidae	*Dicentrarchus labrax* (L.)
	Percidae	*Gymnocephalus cernuus* (L.)
		Perca fluviatilis L.
	Carangidae	*Trachurus trachurus* (L.)
	Mullidae	*Mullus surmuletus* L.
	Sparidae	*Spondyliosoma cantharus* (L.)
	Labridae	*Crenilabrus melops* (L.)
		Ctenolabrus rupestris (L.)
		Labrus berylta Ascanius
	Ammodytidae	*Ammodytes marinus* Raitt
		Ammodytes tobianus L.
		Hyperoplus lanceolatus (Lesauvage)
	Trachinidae	*Trachinus vipera* Cuvier

	Scombridae	*Scombrus scombrus* L.
	Gobiidae	*Aphia minuta* (Risso)
		Gobius forsteri Corbin
		Gobius niger L.
		Gobius paganellus L.
		Pomatoschistus lozanoi (de Buen)
		Pomatoschistus microps (Kröyer)
		Pomatoschistus minutus (Pallas)
		Pomatoschistus pictus (Malm)
	Callionymidae	*Callionymus lyra* L.
	Pholididae	*Pholis gunnellus* (L.)
	Zoarcidae	*Zoarces viviparus* (L.)
	Mugilidae	*Crenimugil labrosus* (Risso)
		Liza auratus (Risso)
		Liza ramada (Risso)
	Atherinidae	*Atherina presbyter* Valenciennes
Scleroparei	Triglidae	*Aspitrigla cuculus* (L.)
		Eutrigla gurnhardus (L.)
		Trigla lucerna L.
		Trigloporus lastoviza (Bonnaterre)
	Cottidae	*Cottus gobio* L.
		Myoxocephalus scorpius (L.)
		Taurulus bubalis (Euphrasen)
		Taurulus lilljeborgi (Collett)
	Agonidae	*Agonus cataphractus* (L.)
	Cyclopteridae	*Cyclopterus lumpus* L.
	Liparidae	*Liparis liparis* (L.)
		Liparis montagui (Donovan)
	Gasterosteidae	*Gasterosteus aculeatus* L.
		Pungitius pungitius (L.)
		Spinachia spinachia (L.)
Heterosomata	Bothidae	*Arnoglossus laterna* (Walbaum)
		Phrynorhombus regius (Bonnaterre)
		Scopthalamus maximus (L.)
		Scopthalamus rhombus (L.)
	Pleuronectidae	*Hippoglossoides platessoides* (Fabricus)
		Limanda limanda (L.)
		Microstomus kitt (Walbaum)
		Platichthys flesus (L.)
		Pleuronectes platessa L.
	Soleidae	*Buglossidium luteum* (Risso)
		Solea solea (L.)
Plectognathi	Balistidae	*Balistes carolinensis* (Gmelin)

	Molidae	*Mola mola* (L.)
Pediculati	Lophiidae	*Lophius piscatorius* L.

References

Adam, P. (1976) Plant sociology and habitat factors in British saltmarshes. PhD thesis, University of Cambridge.

Adam, P. (1978) Geographical variation in British saltmarsh vegetation. *Journal of Ecology*, **66**, 339–366.

Adam, P. (1981) The vegetation of British Saltmarshes. *New Phytologist*, **88**, 143–196.

Adam, P. (1990) *Saltmarsh Ecology*, Cambridge University Press, Cambridge.

Allen, J.R.L. and Pye, K. (1992) *Saltmarshes: Morphodynamics, Conservation and Engineering Significance*, Cambridge University Press, Cambridge.

Ambler, J.W., Cloern, J.E. and Hutchinson, A. (1985) Seasonal cycles of zooplankton from San Francisco bay. *Hydrobiologia*, **129**, 177–197.

Anders, K. (1989) A herpesvirus associated with an epizootic epidermal papillomatosis in European smelt (*Osmerus eperlanus*) In: Ahne, W. and Kurstak E. (eds), *Viruses of the Lower Vertebrates*, pp. 184–197, Springer, Berlin, Heidelberg.

Anders, K. and Möller, H. (1985) Spawning papillomatosis of smelt, *Osmerus eperlanus* L., from the Elbe estuary. *Journal of Fish Diseases*, **8**, 233–235.

Andrews, M.J. (1977) Observations on the fauna of the Metropolitan River Thames during the drought of 1976. *London Naturalist*, 56, 44–56.

Andrews, M.J. (1984) Thames Estuary: pollution and recovery. In: Sheehan, P.J., Miller, D.R., Butler, G.C. and Bourdeau, Ph. (eds), *Effects of Pollutants at the Ecosystem Level*, pp. 195–227, John Wiley and Sons Ltd.

Andrews, M.J. and Rickard, D.G. (1980) Rehabilitation of the Inner Thames Estuary. *Marine Pollution Bulletin*, **11**(11), 327–332.

Andrews, M.J., Aston, K.F.A., Rickard, D.G. and Steel, J.E. (1982) The macrofauna of the Thames Estuary. *London Naturalist*, **61**, 30–62.

Andrews, M.J., Steel, J.E.C. and Cockburn, A.G. (1983) Biological considerations in the setting of quality standards for the tidal Thames. *Water Pollution Control*, **1983**, 52–60.

Anon. (1986) *Salmon Rehabilitation Scheme: Phase 1 Review*, Thames Water Authority internal report.

Arthur, D.R. (1972) Summarising review. In: Barnes, R.S.K. and Green, J. (eds), *The Estuarine Environment*, Academic Science Publishers, London, 133 pp.

Aston, K.F.A. and Andrews, M.J. (1978) Freshwater macroinvertebrates in London's rivers, 1970–1977. *London Naturalist*, **57**, 34–52.

Attrill, M.J. (1992) Thames Estuary Benthic Programme. A site by site report of the results of macroinvertebrate surveys undertaken during 1990–91. *NRATR Biology Report*, 82 pp.

Attrill, M.J. and Thomas, R.M. (1995) Heavy metal concentrations in sediment from the Thames Estuary, UK. *Marine Pollution Bulletin*, **30**, 742–744.

Attrill, M.J. and Thomas, R.M. (1996) Long-term distribution patterns of mobile estuarine invertebrates (Ctenophora, Cnidaria, Crustacea: Decapoda) in relation to hydrological parameters. *Marine Ecology – Progress Series*, **143**, 25–36.

Attrill, M.J., Rundle, S.D. and Thomas, R.M. (1996a) The influence of drought-induced low freshwater flow on an upper-estuarine macroinvertebrate community. *Water Research*, **30**, 261–268.

Attrill, M.J., Ramsay, P.M., Thomas, R.M. and Trett, M.W. (1996b) An estuarine biodiversity hot-spot. *Journal of the Marine Biological Association, UK*, **76**, 161–175.

Barlow, J.P. (1955) Physical and biological processes determining the distribution of zooplankton in a tidal estuary. *Biological Bulletin*, **109**, 211–225.

Barrett, J. and Butterworth, P.E. (1973) The carotenoid pigments of six species of adult Acanthocephala. *Experimentia*, **29**, 651–653.

Barthel, K.-G. (1983) Food uptake and growth efficiency of *Eurytemora affinis* (Copepoda: Calanoida). *Marine Biology*, **74**, 269–274.

Barton, N.J. (1962) *The Lost Rivers of London*, Historical Publications, London.

Beeftink, W.G. (1975) The ecological significance of embankment and drainage with respect to the vegetation of south west Netherlands. *Journal of Ecology*, **63**, 423–458.

Beeftink, W.G. (1977) The coastal salt marshes of western and northern Europe: an ecological and phytosociological approach. In: Chapman, V.J. (ed.), *Wet Coastal Ecosystems*, Elsevier Scientific, Amsterdam.

Birtwell, I.K. (1972) Ecophysiological aspects of tubificids in the Thames Estuary. PhD thesis, University of London.

Birtwell, I.K and Arthur, D.R. (1980) The ecology of tubificids in the Thames Estuary with particular reference to *Tubifex costatus* (Claparede) In: Brinkhurst, R.O. (ed.), *Aquatic Oligochaete Biology*, pp. 331–381, Plenum Press, London.

Blaber, S.J.M. and Blaber, T.G. (1980) Factors affecting the distribution of juvenile and estuarine fish. *Journal of Fish Biology*, **17**, 143–162

Blaxter, J.H.S. (1990) The herring. *Biologist*, **37**, 27–31.

Boorman, L.A. and Ranwell, D.S. (1977) *Ecology of Maplin Sands and the Coastal Zones of Suffolk, Essex and North Kent*, Institute of Terrestrial Ecology, Cambridge.

Bowen, A.J. and Pinless, S.J. (1977) The response of an estuary to the closure of a mobile barrier: Richmond barrier on the Upper Thames estuary. *Estuarine Coastal and Marine Science*, **5**(2), 197–208.

Breslaeuer, T. (1916) Zur Kenntnis der Epidermoidalgeschwulste von Kaltblutern. Histologische Veranderungen des Integuments und der Mundschleimhaut beim Stint (*Osmerus eperlanus* L.) *Archivs Mikroskopische Anatomie*, **87**, 200–263.

Brinkhurst, R.O. (1971) *A Guide to the Identification of British Aquatic Oligochaeta*, FBA Scientific Publication No. 22, FBA, Ambleside, Cumbria, 55 pp.

Burd, F. (1989) *The Saltmarsh Survey of Great Britain. An Inventory of British Saltmarshes*, Research and Survey in Nature Conservation, No. 17, Nature Conservancy Council, Peterborough.

Burd, F. (1992) *Erosion and Vegetation Change on the Saltmarshes of Essex and North Kent between 1973 and 1988*, Research and Survey in Nature Conservation, No. 42, Nature Conservancy Council, Peterborough.

Burkill, P.H. (1983) Sampling and analysis of zooplankton populations. In: Morris, A.W. (ed.), *Practical Procedures for Estuarine Studies*, pp. 139–184, The Natural Research Council, Plymouth.

Burkill, P.H. and Kendall, T.F. (1982) Production of the copepod, *Eurytemora affinis* in the Bristol channel. *Marine Ecology – Progress Series*, **7**, 21–31.

Busch, A. and Brenning, U. (1992) Studies on the status of *Eurytemora affinis* (Poppe, 1880) (Copepod, Calanoida). *Crustaceana*, **62**(1), 13–38.

Carter, N. (1933a) A comparative study of the algal flora of two salt marshes. Part II. *Journal of Ecology*, **21**, 128–208.

Carter, N. (1933b) A comparative study of the algal flora of two salt marshes. Part III. *Journal of Ecology*, **21**, 385–403.

Carter, R.W.G. (1988) *Coastal Ecology, An Introduction to the Physical, Ecological and Cultural Systems of Coastlines*, Academic Press, London.
Chapman, V.J. (1934) The ecology of Scolt Head Island. In: Steers, J.A. (ed.), *Scolt Head Island*, pp. 77–145, Heffer, Cambridge.
Chapman, V.J. (1941) Studies in salt-marsh ecology. Section VIII. *Journal of Ecology*, **29**, 69–82.
Chapman, V.J. (1964) *Coastal Vegetation*, Pergamon Press, Oxford.
Chapman, V.J. (1976) *Coastal Vegetation*, 2nd edn, Pergamon Press, Oxford.
Charman, K. (1977) The grazing of *Zostera* by wildfowl in Britain. *Aquaculture*, **12**, 229–233.
Charman, K., Fojt, W. and Penny, S. (1986) *Saltmarsh Survey of Great Britain: Bibliography*, Research and Survey in Nature Conservation No. 3, Nature Conservancy Council, Peterborough.
Christensen, T. (1987) *Seaweeds of the British Isles, Volume 4: Tribophyceae (Xanthophyceae)* . British Museum (Natural History), London.
Claridge, P.N., Potter, I.C. and Hardisty, M.W. (1986) Seasonal changes in movements, abundance, size composition and diversity of the fauna of the Severn estuary. *Journal of the Marine Biological Association, UK*, **66**, 229–258.
Clarke, K.R. and Warwick, R.M. (1994) *Change in Marine Communities: an Approach to Statistical Analysis and Interpretation*, Natural Environmental Research Council, UK, 144 pp.
Collins, N.R. and Williams, R. (1981) Zooplankton of the Bristol channel and Severn estuary. The distribution of four copepods in relation to salinity. *Marine Biology*, **64**, 273–283.
Collins, N.R. and Williams, R. (1982) Zooplankton communities in the Bristol channel and the Severn estuary. *Marine Ecology – Progress Series*, **9**, 1–11.
Cushing, D.H. (1986) The migration of larval and juvenile fish from spawning ground to nursery ground. *Journal Conseil Internationale de l'Exploration de la Mer*, **43**, 43–49.
Dalby, D.H. (1970) The salt marshes of Milford Haven, Pembrokshire. *Field Studies*, **3**, 297–330.
Davidson, N.C. (1990) The Conservation of British North Sea Estuaries. *Hydrobiologia*, **195**, 145–162.
Davidson, N.C., Laffoley, D.d'A., Doody, J.P. *et al.* (1991) *Nature Conservation and Estuaries in Great Britain*, Nature Conservancy Council, Peterborough.
Davies, J.F. (1969) Estuaries as economic growth points in Britain. *Koninklijk Nederlands Aardrijkskundig Genootschap Geografisch Tijschrift*, **III**(1), 22–35.
Dawson, J.K. (1974) Copepods (Arthropoda: Crustacea: Copepoda). In: Hart, C.W. Jr and Fuller, S.L.H. (eds), *Pollution Ecology of Estuarine Invertebrates*, pp. 145–170, Academic Press, New York.
Den Hartog, C. (1960) Comments on the Venice-system for the classification of brackish waters. *Internationale Revue der Gesamten Hydrobiologie*, **45**, 481–485.
Den Hartog, C. (1983) Structural uniformity and diversity in *Zostera*-dominated communities in Western Europe. *Marine Technology Society Journal*, **17**, 6–14.
Den Hartog, C. (1987) 'Wasting disease' and other dynamic phenomena in *Zostera* beds. *Aquatic Botany*, **27**, 3–14.
Dillenius, J.J. (1742) *Historia Muscorum*, Oxonii.
Dillenius, J.J. and Ray, J. (1724) *Synopsis Methodica Stirpium Britannicarum*, edn 3, London.
Doxat, J. (1977) *The Living Thames: the Restoration of a Great Tidal River*, Hutchinson Benham, London, 96 pp.
Dring, M.J. (1984) Mean irradiance levels at different heights in the intertidal and upper subtidal zones: a general model and its application to shores in the Bristol Channel. *British Phycological Journal*, **19**, 192.

Dugan, P.J. (ed.) (1990) *Wetland Conservation: a Review of Current Issues and Actions Required*, World Conservation Union, Gland, Switzerland.
Ekins, R. (1990) *Changes in the Extent of Grazing Marshes in the Greater Thames Estuary*. Internal Report produced by the Royal Society for the Protection of Birds, Sandy, Bedfordshire, UK.
Elliot, J.M. and Mann, K.H. (1979) *A Key to the British Freshwater Leeches*, FBA Scientific Publication No. 40, FBA, Ambleside, Cumbria.
Elliott, M., O'Reilly, M.G. and Taylor, C.J.L. (1990) The Forth estuary: a nursery and overwintering area for North Sea fishes. *Hydrobiologia*, **195**, 89–103.
El-Maghraby, A.M.A. (1956) The inshore plankton of the Thames estuary. PhD thesis, University of London, 238 pp.
Eppy, D.R. (1989) *A Survey of Water Quality in the Tidal River Thames and the River Lee Using the Chironomid Pupal Exuviae Technique (CPET)*, Thames Water Authority (NRA Unit) Biology Report, 12 pp.
Evans, H.M. (1936) *A Short History of the Thames Estuary*, Imray, Laurine, Norie and Wilson, London, 78 pp.
Fojt, W. (1978) Tidal effects on salt marsh vegetation at Faversham Creek, Kent. *Kent Field Club Transactions*, **6**, 171–179.
Fulton, R.S. (1982) Preliminary results of an experimental study of the effects on mysid predation on estuarine zooplankton community structure. *Hydrobiologia*, **93**, 79–84.
Fulton, R.S. (1984) Distribution and community structure of estuarine copepods. *Estuaries*, **7**, 38–50.
Gaedke, U. (1990) Population dynamics of the calanoid copepods *Eurytemora affinis* and *Acartia tonsa* in the Ems-Dollart estuary: a numerical simulation. *Archives für Hydrobiologie*, **118**, 185–226.
Gameson, A.L.H. and Johnson, N.H. (1964) Effects of polluting discharges on the Thames estuary. *Technical Paper on Water Pollution Research*, **11**, i–xxvii + 1–609.
Gameson, A.L.H. and Wheeler, A. (1977) Restoration and recovery of the Thames Estuary. In: Cairns, J. (ed.), *Recovery and Restoration of Damaged Ecosystems*, pp. 72–101, University Press of Virginia, Charlottesville.
Gee, J.M. (1961) Ecological studies in South Benfleet Creek with special reference to *Corophium*. *Essex Naturalist*, **6**, 291–308.
Gibbs, P.E. (1993) A male genital defect in the dog-whelk, *Nucella lapillus* (Neogastropoda), favoring survival in a Tbt-polluted area. *Journal of the Marine Biological Association, UK*, **73**, 667–678
Gilbert, S. and Horner, R. (1992) *The Thames Barrier*, Thomas Telford Ltd, London.
Glenny, C.M. and Kinniburgh, J.H. (1991) Low flow management of the Lower Thames. *Proceedings of the Third National Hydrology Symposium*, British Hydrological Society, **2**, 29–36.
Goodman, P.J., Braybrooks, E.M. and Lambert, J.M. (1959) Investigations into die-back in *Spartina townsendii*. I. The present status of *S. townsendii* in Britain. *Journal of Ecology*, **47**, 651–677.
Gordon, C., Bark, A.W. and Bailey, R.G. (1993) The effect of fibre on the feeding behaviour and efficiency of estuarine plankton in the Thames tideway. *Verh. int. Verein. Limnol.* **25**,
Gough, P. (1990) Rebirth of the Thames salmon. *Salmon, Trout and Sea-Trout*, **August**, 60–63.
Gourlay, K.A. (1992) *World of Waste: Dilemmas of Industrial Development*, Zed Books, London.
Gray, A.J. (1971) Variation in *Aster tripolium* with particular reference to some British population. PhD thesis, Keele University.
Gray, A.J. (1974) The genecology of saltmarsh plants. *Hydrobiology Bulletin*, **8**, 152–165.

Griffiths, I.M. and Lloyd, P.J. (1985) Mobile oxygenation in the Thames Estuary. *Effluent and Water Treatment Journal*, **5**, 165–169.
Grubb, M., Koch, M., Thomson, K. *et al.* (1993) *The 'Earth Summit' Agreements: a Guide and Assessment*, Earthscan, London.
Haedrich, R.L. (1983) Estuarine fishes. In: Ketchum, B. (ed.), *Estuaries and Enclosed Seas*, pp. 183–207, Elsevier, Amsterdam.
Harding, D., Nichols, J.H. and Tungate, D.S. (1978) The spawning of plaice (*Pleuronectes platessa*) in the southern North Sea and English Channel. *Rapp. P.-v Reun. Cons. int. Explor. Mer*, **172**, 102–113.
Harmsworth, G.C. and Long, S.P. (1986) Assessment of saltmarsh erosion in Essex, England, with reference to the Dengie Peninsula. *Biological Conservation*, **35**, 377–387.
Harrison, J. and Grant, P. (1976) *The Thames Transformed: London's River and its Waterfowl*, Andre Deutsch, London, 240 pp.
Heckman, C.W. (1986) The anadromous migration of a calanoid copepod, *Eurytemora affinis* (Poppe, 1880) in the Elbe estuary. *Crustaceana*, **50**, 176–181.
Heerkloss, R., Brenning, U., Ihienfeld, R. and Frank, R. (1990) Influence of temperature and epizoic ciliates on the growth of *Eurytemora affinis* (Poppe) (Calanoida, Copepoda) under laboratory conditions. *Wiss. Z. Universitat Rostock N-Reihe*, **39**, S.12–15.
Heinle, D.R. and Flemer, D.A. (1975) Carbon requirements of a population of the estuarine copepod *Eurytemora affinis*. *Marine Biology*, **31**, 235–247.
Heinle, D.R., Harris, R.P., Ustach, J.F. and Flemer, D.A. (1977) Detritus as food for estuarine copepods. *Marine Biology*, **40**, 341–353.
Herbert, A.P. (1966) *The Thames*, Weidenfeld and Nicolson, London.
Herman, P.M.J. and Scholten, H. (1990) Can suspension-feeders stabilise estuarine ecosystems? In: Barnes, M. and Gibson, R.N. (eds), *Trophic Relationships in the Marine Environment*, pp. 104–116, (Proceedings 24th European Marine Biological Symposium), Aberdeen University Press, Aberdeen.
Hine, P.M. and Kennedy, C.R. (1974) Observations on the distribution, specificity and pathogenicity of the acanthocephalan *Pomphorhynchus laevis* (Müller). *Journal of Fish Biology*, **6**, 521–535.
HMSO (1964) *Effects of Polluting Discharges on the Tidal Thames*, Water Pollution Research Laboratory, Technical Paper 11, HMSO, London.
HMSO (1992) *Coastal Zone Protection and Planning, Second Report. House of Commons Environment Committee*, HMSO, London.
Hough, A.R. and Naylor, E. (1992) Distribution and postion maintenance of the estuarine mysid *Neomysis integer*. *Journal of the Marine Biological Association, UK*, **72**, 869–876.
Houghton, J.T. *et al.* (eds) (1996) *Climate Change 1995. The Science of Climate Change*, published for the Inter-Governmental Panel on Climate Change by Cambridge University Press, Cambridge.
Huddart, R. and Arthur, D.R. (1971a) Shrimps and whitebait in the polluted Thames estuary. *International Journal of Environmental Studies*, **2**, 21–34.
Huddart, R. and Arthur, D.R. (1971b) Lampreys and teleost fish other than whitebait in the polluted Thames Estuary. *International Journal of Environmental Studies*, **2**, 143–152.
Huddart, R. and Arthur, D.R. (1971c) Shrimps in relation to oxygen depletion and its ecological significance in a polluted estuary. *Environmental Pollution*, **2**, 13–35.
Hunter, J. and Arthur, D.R. (1978) Some aspects of the ecology of *Peloscolex benedeni* Udekem (Oligochaeta Tubificidae) in the Thames estuary. *Estuarine and Coastal Marine Science*, **6**, 197–208.

Hussey, A. and Long, S.P. (1982) Seasonal changes in weight of above- and below-ground vegetation and dead plant material in a salt marsh at Colne Point, Essex. *Journal of Ecology*, **70**, 757–772.

Hutchinson, P. (1983) The ecology of smelt, *Osmerus eperlanus*, from the River Thames and the River Cree. PhD thesis, University of Edinburgh.

IMER (1984a) *Predicted Effects of Proposed Changes in Patterns of Water Abstraction on the Ecosystems of the Lower River Thames and its Tidal Estuary: Final Report*, Institute for Marine Environmental Research Miscellaneous Publications (9), IMER, Plymouth, Devon, 229 pp.

IMER (1984b) *Predicted Effects of Proposed Changes in Patterns of Water Abstraction on the Ecosystems of the Lower River Thames and its Tidal Estuary: Bibliography*, Institute for Marine Environmental Research Miscellaneous Publications (10), IMER, Plymouth, Devon, 86 pp.

Inglis, C.C. and Allen, F.H. (1957) The regimen of the Thames Estuary as affected by currents, salinities, and river flow. *Proceedings of the Institute of Civil Engineers*, **7**, 827–868

Ingrouille, M.J. and Pearson, J. (1987) The pattern of morphological variation in the *Salicornia europaea* L. aggregate (Chenopodiaceae). *Watsonia*, **16**, 269–281.

Jefferies, H.P. (1967) Saturation of estuarine communities by congeneric associates. In: Lauff, G.E. (ed.), *Estuaries*, pp. 500–508, American Association for the Advancement of Science 83, Washington DC.

John, D.M. and Moore, J.A. (1985a) Observations on the phytobenthos of the freshwater Thames I. The environment, floristic composition and distribution of the macrophytes (principally macroalgae) *Archiv für Hydrobiologie*, **102**, 435–459.

John, D.M. and Moore, J.A. (1985b) Observations on the phytobenthos of the freshwater Thames II. The floristic composition and distribution of the smaller algae sampled using artificial surfaces. *Archiv für Hydrobiologie*, **103**, 83–97.

John, D.M., Johnson, L.R. and Moore, J.A. (1989a) Observations on *Thorea ramosissima* Bory (Batrachospermales, Thoraceae), a freshwater red alga rarely recorded in the British Isles. *British Phycological Journal*, **24**, 90–102.

John, D.M., Johnson, L.R. and Moore, J.A. (1989b) The red alga *Thorea ramosissima*: its distribution and status in the Thames catchment. *The London Naturalist*, **68**, 49–53.

John, D.M., Johnson, L.R. and Moore, J.A. (1990) Observations on the phytobenthos of the freshwater Thames III. The floristic composition and seasonality of algae in the tidal and non-tidal river. *Archiv für Hydrobiologie*, **103**, 83–97.

Juggins, S. (1992) Diatoms in the Thames estuary, England: ecology, palaeoecology, and salinity transfer function. *Bibliotheca Diatomologica*, **25**, 1–216.

Kabata, Z. (1970) Crustacea as enemies of fishes. In: Snieszko, S.F. and Axelrod, H.R. (eds) *Diseases of Fishes; Book 1*, T.F.H. Publications, New Jersey.

Kennedy, C.R. (1984) The status of flounders, *Platichthys flesus* L., as hosts of the acanthocephalan *Pomphorhynchus laevis* (Müller) and its survival in marine conditions. *Journal of Fish Biology*, **24**, 135–149.

Kennedy, C.R. (1985) Regulation and dynamics of acanthocephalan populations. In: Crompton, D.W.T. and Nickol, B.B. (eds), *The Biology of the Acanthocephala*, pp. 385–416, Cambridge University Press, Cambridge.

Kennedy, C.R., Broughton, P.F. and Hine, P.M. (1978) The status of brown and rainbow trout, *Salmo trutta* and *S. gairdneri* as hosts of the acanthocephalan *Pomphorhynchus laevis*. *Journal of Fish Biology*, **13**, 265–275.

Kennedy, C.R., Bates, R.M. and Brown, A.F. (1989) Discontinuous distributions of the fish acanthocephalans *Pomphorhynchus laevis* and *Acanthocephalus anguillae* in Britain and Ireland: an hypothesis. *Journal of Fish Biology*, **34**, 607–619.

Kremer, P. and Kremer, P.N. (1988) Energetic and pulsed food availability for zooplankton. *Bulletin of Marine Science*, **43**, 797–809.

Kuipers, B.R., Gaedke, U., Enserink, L. and Witte, H. (1990) Effect of ctenophore predation on mesozooplankton during a spring outbreak of *Pleurobrachia pileus*. *Netherlands Journal of Sea Research*, **26**, 111–124.

Lambshead, P.J.D. and Paterson, G.L.J. (1986) Ecological analysis – an investigation of numerical cladistics as a method for analysing ecological data. *Journal of Natural History*, **20**, 895–909.

Lauff, S.M. (1967) *Estuaries*, Publication No. 83, American Association for the Advancement of Science, Washington DC.

Leach, S.J. (1988) Rediscovery of *Halimione pedunculate* (L.) Aellen in Britain. *Watsonia*, **17**, 170–171.

Lee, S. and Whitfield, P.J. (1992) Virus-associated spawning papillomatosis in smelt, *Osmerus eperlanus* L., in the River Thames. *Journal of Fish Biology*, **46**, 503–510.

Lincoln, R.J. (1979) *British Marine Amphipoda: Gammaridea*, British Museum (Natural History) publication No. 818, London, 658 pp.

Lloyd, P.J. and Cockburn, A.G. (1983) Pollution management and the tidal Thames. *Water Pollution Control*, **1983**, 392–401.

Lloyd, P.J. and Whiteland, M.R. (1990) Recent developments in oxygenation of the tidal Thames. *Journal of the Institution of Water and Environmental Management*, **4**, 103–111.

Long, S.P. and Mason, C.F. (1983) *Saltmarsh Ecology*, Blackie, Glasgow.

Long, S.P. and Woolhouse, H.W. (1979) Primary production in *Spartina* marshes. In: Jefferies, R.L. and Davy, A.J. (eds), *Ecological Processes in Coastal Environments*, pp. 333–352, Blackwell Scientific, Oxford.

Lumkin, P.A. (1971) Plankton and pollution in the Thames Estuary from Barking Reach to Barrow Deep (S.N. Sea). MPhil thesis, University of London, 457 pp.

Macan, T.T. (1969) *A Key to the British Fresh- and Brackish-water Gastropoda*, 3rd edn, FBA Scientific Publication No.13, FBA, Ambleside, Cumbria, 44 pp.

Mackenzie, K. and Gibson D.I. (1970) Ecological studies of some parasites of plaice *Pleuronectes platessa* L. and flounder *Platichthys flesus* (L.) *Symposia of the British Society of Parasitology*, **8**, 1–42.

Malloch, A.J.C. (1988) *Vespan II. A computer package to handle and analyse multivariate species data and handle and species distribution data*, Lancaster University, Lancaster.

Marchant, C.J. (1968) Evolution in *Spartina* (Graminae) I. The history and morphology of the genus in Britain. 2. Chromosomes, basic relationships and the problems. *Linnean Society Journal, Botany*, **60**, 381–409.

Meng, L. and Orsi, J.J. (1991) Selective feeding by larval striped bass on native and introduced copepods. *Transactions of the American Fisheries Society*, **120**, 187–192.

Miller, P.J. (1963) Studies on the biology and taxonomy of the British Gobiid fishes. PhD thesis, University of Liverpool.

Ministry of Housing and Local Government (1961) *Pollution of the Tidal Thames*, HMSO, London.

Möller, H. (1987) Pollution and parasitism in the aquatic environment. *International Journal of Parasitology*, **17**, 353–361.

Möller, H. and Scholz, U. (1991) Avoidance of oxygen-poor zones by fish in the Elbe estuary. *Journal of Applied Ichthyology*, **7**, 176–182.

Mowah, V. (1991) Diet of Thames estuary fish. Final year project, Division of Biotechnology, Polytechnic on Central London, 105 pp.

Moyle, P.B., Herbold, B., Stevens, D.E. and Miller, L.W. (1992) Life history and status of delta smelt in the Sacramento–San Joaquin Estuary, California. *Transactions of the American Fisheries Society*, **121**, 67–77.

Mueller-Dombios, D. and Ellenberg, H. (1974) *Aims and Methods of Vegetation Ecology*, Wiley, New York.

Munro, M.A. (1992) Studies on the estuarine strain of *Pomphorhynchus laevis* (Acanthocephala) in the Thames estuary. PhD thesis, University of London.

Munro, M.A., Whitfield, P.J. and Diffley, R. (1989) *Pomphorhynchus laevis* (Müller) in the flounder *Platichthys flesus* L. in the tidal River Thames: population structure, microhabitat utilisation and reproductive status in the field and under conditions of controlled salinity. *Journal of Fish Biology*, **35**, 719–735.

Munro, M.A., Reid, A. and Whitfield, P.J. (1990) Genomic divergence in the ecologically differentiated English freshwater and marine strains of *Pomphorhynchus laevis* (Acanthocephala: Palaeacanthocephala): a preliminary investigation. *Parasitology*, *101*, 451–454.

Myers, J.E. (1954) Survey and comparison of the natural and inner salt marsh at Leigh-on-Sea, Essex. *Essex Naturalist*, **29**, 155–175.

National Water Council (1976) *Report of the Working Party on Consent Conditions for Effluent Discharges to Freshwater Streams*, HMSO, London.

Nickol, B.B. (1985) Epizootiology. In: Crompton, D.W.T. and Nickol, B.B. (eds), *The Biology of the Acanthocephala*, pp. 307–346, Cambridge University Press, Cambridge.

NRC (1994) *Restoring and Protecting Marine Habitats*, National Academy Press, Washington DC.

Odd, N.V.M. (1988) Mathematical modelling of mud transport in estuaries. In: Dronkes, J. and van Leusen, W. (eds), *Physical Processes in Estuaries*, pp. 503–531, Springer-Verlag, Berlin.

Orsi, J.J. and Mecum, W.L. (1986) Zooplankton distribution and abundance inthe Sacramento–San Joaquin delta in relation to certain environmental factors. *Estuaries*, **9**, 326–339.

Othman, S.B. (1980) The distribution of salt marsh plants and its relation to edaphic factors with particular reference to *Puccinellia maritima* and *Spartina townsendii*. PhD thesis, University of Essex, Colchester.

Pearce, F. (1992) Time to sound the retreat for sea defences. *New Scientist*, **1843**, 5.

Pickett, G.D. (1989) The sea-bass. *Biologist*, **36**, 89–95.

Pielou, E.C. (1966) The measurement of diversity in different types of biological collections. *Journal of Theoretical Biology*, **13**, 131–144.

Pirazzoli, P.A. (1986) Secular trends of relative sea-level (rsl) changes indicated by tide-gauge records. *Journal of Coastal Research*, NS **11**, 1–26.

Potter, J.H. (1971) Pollution and its control in the tidal Thames. *Community Health*, **3**, 103–110.

Powell, M.D. and Berry, A.J. (1990) Ingestion and regurgitation of living and inert materials by the estuarine copepod *Eurytemora affinis* and the influence of salinity. *Estuarine, Coastal and Shelf Science*, **31**, 763–773.

Preddy, J.E. (1954) The mixing and movement of water in the estuary of the Thames. *Journal of the Marine Biological Association, UK*, **33**, 645–662.

Price, J.H. (1982) *Fucus* and *Ulva* recolonisation along the Thames. *The London Naturalist*, **61**, 71.

Price, J.H. (1983) The distribution of benthic macroalgae along the inner saline reaches of the tidal Thames: recent changes. *Transactions of the Kent Field Club*, **9**, 65–69.

Price, S.M. and Price, J.H. (1983) The River Wandle: studies on the distribution of aquatic plants. *The London Naturalist*, **62**, 26–58.

Pritchard, D.W. (1955) Estuarine circulation patterns. *American Society of Civil Engineers*, **81**, 717/1–717/11.

Randerson, P.F. (1979) A simulation model of salt-marsh development and plant ecology. In: Knights, B. and Phillips, A.J. (eds), *Estuarine and Coastal Land Reclamation and Water Storage*, Saxon House, UK.

Ranwell, D.S. (1972) *Ecology of Salt Marshes and Sand Dunes*, Chapman & Hall, London.
Rassai, M. (1992) Studies on the parasites of eels (*Anguilla anguilla* L.) in the Thames catchment, with emphasis on salinity-related infection patterns and parasite pathogenesis. PhD thesis, University of London.
Rice, C.H. (1938) Studies in the phytoplankton of the River Thames, (1928–1932) Parts I and II. *Annals of Botany N.S.*, **2**, 539–557, 559–581.
Riley, J.D., Symonds, D.L. and Woolner, L. (1981) On the factors influencing the distribution of 0-group demersal fish in coastal waters. *Rapp.P.-v Reun. Cons. int. Explor. Mer*, **178**, 223–228.
Riley, J.P. and Chesters, R. (1971) *Introduction to Marine Chemistry*, Academic Press, London and New York.
Roper, F.C.S. (1954) Some observations on the Diatomaceae of the Thames. *Microscopy Journal*, **2**, 67–80.
Royal Commission on Environmental Pollution (1972) *Third Report: Pollution in some British Estuaries and Coastal Waters*, HMSO, London.
Sedgwick, R.L. (1979) Some aspects of the ecology and physiology of Nekton in the Thames estuary, with special reference to the shrimp *Crangon vulgaris*. PhD thesis, University of London, 675 pp.
Sedgwick, R.W. and Arthur, D.R. (1979) A natural pollution experiment; the effects of a sewage strike on the fauna of the Thames estuary. *Environmental Pollution*, **11**, 137–160.
Sellner, K.G. and Bundy, M.H. (1987), Preliminary results of experiments to determine the effects of suspended sediments on the estuarine copepod *Eurytemora affinis*. *Continental Shelf Science*, **7**, 1435–1438.
Sexton, J.R. (1988) Regulation of the River Thames: a case study on the Teddington Flow Proposal. *Regulated Rivers: Research and Management*, **2**, 323–333.
Soltanpour-Gargari, A. and Wellershaus, S. (1984) *Eurytemora affinis* – the estuarine plankton copepod in the Weser. *Veroff. Inst. Meeresforsch. Bremerh.*, **20**, 103–117.
Soltanpour-Gargari, A. and Wellershaus, S. (1987) Very low salinity stretches in estuaries – the main habitat of *Eurytemora affinis*, a planktonic copepod. *Veroff. Inst. Meeresforsch. Bremerh.*, **31**, 199–208.
Stace, C.A. (1991) *New Flora of the British Isles*, Cambridge University Press, Cambridge.
Swafford, D.L. (1985) *Phylogenetic Analysis Using Parsimony*, PC version 2.4.
Taylor, C.J.L. (1987) The zooplankton of the Forth, Scotland. *Proceedings of the Royal Society of Edinburgh*, **93B**, 377–388.
Thompson, T.E. and Brown, G.H. (1976) *British Opisthobranch Molluscs*, Synopses of the Fauna of the British Isles, No. 8, Academic Press, London.
Tittley, I. (1985a) Zonation and seasonality of estuarine benthic algae: artificial embankments in the River Thames. *Botanica Marina*, **28**, 1–8.
Tittley, I. (1985b) Seaweed communities on the artificial coastline of southeastern England. 1: Reclaimed saline wetland and estuaries. *Transaction of the Suffolk Naturalists' Society*, **21**, 54–64.
Tittley, I. (1986) Seaweed communities on the artificial coastline of southeastern England. 2: open sea shores. *Transactions of the Kent Field Club*, **10**, 55–67.
Tittley, I. and Price, J.H. (1977a) The marine algae of the tidal Thames. *The London Naturalist*, **56**, 10–17.
Tittley, I. and Price, J.H. (1977b) An atlas of the seaweeds of Kent. *Transactions of the Kent Field Club*, **7**, 1–80.
Tittley, I., Fletcher, R.L. and Price, J.H. (1985) Additions to an atlas of the seaweeds of Kent. *Transactions of the Kent Field Club*, **10**, 3–11.
Trett, M.W., Feil, R.L. and Forster, S.J. (1990) *Meiofaunal Assemblages of the Thames Estuary, Module II*, Report to the National Rivers Authority Thames Region, 8 pp.

Van der Veer, H.W. (1985) Impact of coelenterate predation on larval plaice, *Pleuronectes platessa*, and flounder, *Platichthys flesus*, stock in the western Wadden Sea. *Marine Ecology – Progress Series*, **25**, 229–238.

Vernberg, W.B. and Vernberg, F.J. (1976) Physiological adaptations of estuarine animals. *Oceanus*, **19**, 48–54.

Vuorinen, I. (1987) Vertical migration of *Eurytemora* (Crustacea, Copepoda): a compromise between the risks of predation and decreased fecundity. *Journal of Plankton Research*, **9**, 1037–1046.

Walker, P. and Davies, I.L. (1986) *The Lowestoft Frame Trawl*, Fisheries Research Technical Report No. 81, MAFF, Lowestoft.

Water Pollution Research Laboratory (1964) *Effects of Polluting Discharges on the Tidal Thames*, Water Pollution Research Technical Paper 11, HMSO, London.

Wells, A.L. (1938) Some notes on the plankton of the Thames estuary. *Journal of Animal Ecology*, **7**, 105–124.

Wharfe, J.R. (1977) An ecological survey of the benthic invertebrate macrofauna of the lower Medway estuary, Kent. *Journal of Animal Ecology*, **46**, 93–113.

Wheeler, A. (1958) The Fishes of the London Area. *London Naturalist*, **1957**, 80–101.

Wheeler, A. (1969a) Fish life and pollution in the lower Thames: A review and preliminary report. *Biological Conservation*, **2**, 25–30

Wheeler, A.C. (1969b) *The Fishes of the British Isles and North West Europe*, Macmillan, London.

Wheeler, A. (1978) *Key to the Fishes of Northern Europe*, William Clowes and Sons, London, 380 pp.

Wheeler, A.C. (1979) *The Tidal Thames. The History of a River and its Fishes*, Routledge and Kegan Paul, London, 228 pp.

Wheeler, A. (1992) A list of the common and scientific names of fishes of the British Isles. *Journal of Fish Biology*, **41**, Suppl. A.

Whitehead, P.J.P., Bauchot, M.-L., Hureau, J.-C. *et al.* (eds) (1989) *Fishes of the North-eastern Atlantic and the Mediterranean*, 3 vols, UNESCO

Whitfield, P.J., Pilcher, M.W., Grant, H.J. and Riley, J. (1988) Experimental studies on the development of *Lernaeocera branchialis* (Copepoda: Pennellidae): population processes from egg production to maturation on the flatfish host. *Hydrobiologia*, **167/168**, 579–586.

Wood, L.B. (1980) The rehabilitation of the tidal River Thames. *The Journal of the Institute of Public Health Engineers*, **8**, 112–120.

Wood, L.B. (1982) *The Restoration of the Tidal Thames*, Adam Hilger Ltd, Bristol.

Wooster, W.S., Lee, A.J. and Dietrich, G. (1969), article in *Deep Sea Research*, **16**, 321.

Wyer, D.W., Boorman, L.A. and Waters, R. (1977) Studies on the distribution of *Zostera* in the outer Thames Estuary. *Aquaculture*, **12**, 215–227.

Index

Note: page numbers in *italics* refer to tables, those in **bold** refer to figures.

abstraction 32–3
  freshwater flow over Teddington Weir 92
Acanthocephala 141, 142
acanthocephalan parasites
  eel *165*
  flounder 163
  smelt *163*
acanthors, shelled 144, 153
agriculture 192
  arable 20
  nitrate runoff 46
algae 49–51
  benthic habitats 50–1
  blooms 192
  blue-green 55
  colonization
    habitats 65
    substrates 50
  data ordination 59, 62
  distribution 58–9, *60–1*, 62–4
    changes 57–8
    potential limits 64
  diversity 65
  occurrence changes 57–8
  reappraisal 51–3
  recorded sites *60–1*
  saltmarsh 54
  seasonality 55–7
  species
    absence 58
    diversity 54
    list 203–6
    loss 58
amenity value 9, 191–2, 193
amphipods 72
  benthic macroinvertebrate zone 2 94, 95
  benthic macroinvertebrate zone 3 99, 100
  benthic macroinvertebrate zone 4 104, 108
  benthic macroinvertebrate zone 5 111
  see also Gammarus zaddachi
Anas penelope 174
angling 9
Anguilla anguilla 122, *165*
Annelida
  benthic macroinvertebrate zone 4 104, 109–10
  species list 216–19
Aster tripolium (sea aster) 187, 188
atmospheric conditions 34
Atriplex 175
Atriplex pedunculata 188
Atriplex portulacoides 187, 188
Atylus falcatus 111, **112**

bacteria, water temperature 35
bait-digging 103
bank walls
  macroalgae 54
  see also embankments
barrages 193
barriers
  drop-gate 21, 26
  half-tide 7, 10
  see also Thames Barrier
bass **119**, 125, **126**
  abundance 132
  Pomphorhynchus laevis infection 150
  sea 198
beach amenities 191–2
Benfleet and Southend Marshes 25
biochemical oxygen demand (BOD) 41
Biological Monitoring Working Party 23
biological parameters *200*
biomass
  benthic macroinvertebrate zone 4 103, 105, 107
  totals *95*
birds 174–5
  piscivores 2
  wading 189
bivalves
  benthic macroinvertebrate zone 1 90–91
  benthic macroinvertebrate zone 2 95
  benthic macroinvertebrate zone 3 98–9
  benthic macroinvertebrate zone 4 103, 104, 107, 108, 110
  benthic macroinvertebrate zone 5 111
Blackwater river 7
blue-green algae 55
brackish water reaches 53–4
Branta bernicla 174
Bray–Curtis similarity measures **87**

Brent goose grazing 174–5
Bryozoa species list 228–9

calanoid copepods 68, 71
  abundance 72–3
  distribution 72
carbon dioxide, emission reduction 197
carbonaceous oxidation 35
Carcinas maenas 94
catworm 95
Cestoda 141, 142
  eel parasites *165*
Chaetognatha
  benthic macroinvertebrate zone 4 108
  species list 229
Chelicerata species list 224
chemical cycle interrelationships **47**
chironomids 2
chloride concentrations 37, **38**, 39, **40**
Chlorophyta
  distribution 59
  local variation 62
Chordata species list 229–33
ciliates, epizoic 84
cirratulids 107–8, 109
Cladocera 71
Clupea harengus 126–7
Clupeidae 129
Cnidaria 69–70
coal-gas, waste liquors 12
coastal engineering 197
coastal erosion prevention 177
coastal zone development 195–6
Coastal Zone Protection and Planning Report (1992) 195
cockle fishery 9
Coelenterata
  flounder larvae predation 123–4
  species list 207–8
Colne river 7
condition factor (*K*) analysis 155–6, 167
cooling water 8–9
  aeration 117
  discharge 34
  West Thurrock Power Station 116–17, 118
Copepoda 68, 71
  cyclopoid 71, 74, **75**
  predation 83
copepods, parasitic 141, 142
  flounder 152, 164
  whiting 164
copper 47
cord-grass 176
Coriolis forces 77
crab
  Chinese mitten 2, 93
  shore 94
creek habitat 25
Crouch river 7
crustaceans
  benthic macroinvertebrate zone 2 95–6
  benthic macroinvertebrate zone 3 99, 100
  benthic macroinvertebrate zone 4 103–4, 105, 107, 108
  benthic macroinvertebrate zone 5 111
  nauplii mortality 84
  parasitic 141
  species list 219–24
Cryptocotyle lingua 142, 162
ctenophores 69–70
  copepod distribution 83
  flounder larvae predation 124
Cucullanus 163
Cucullanus heterochrous 142
Cucullanus minutus 141–2
currents
  plankton distribution 77
  speed on bends 77
  tidal 30
cyclopoid copepods 71
  distribution 74, **75**
Cyprinidae 129
cystacanth 144, 146, 147

dab, dendrogram 132
dace 23
dams 195
dangerous substances 46–7
daphniids 71
DECCA navigational system 134
DECORANA 52, 59, 62, 63
deforestation 195
dendrogram, fish species 132, **133**
deoxygenation of water 192
depth 27, **28**
detergents, non-biodegradable synthetic 13
detritus 81
developing world 192
diatoms 49–50
  seasonality 55–6
  sediment community 65
  silt-binding 54, 55
Dicentrarchus labrax 125, 150, 198
Digenea 141, 142
  eel parasites *165*
  smelt parasites 162, *163*
diversity, Shannon–Wiener index 121–2
drainage
  management 12
  see also storm drainage
drift-line communities 175, 184

east coast, rate of sinking 172
EC Dangerous Substances in Water Directive 20
Echinodermata
  benthic macroinvertebrate zone 4 105, 107
  benthic macroinvertebrate zone 5 111
  species list 229

ecocladistic analysis 63–4
ecocladograms 52, 53, 62–3, **64**
eel 122
  helminth parasites 164, *165*
eelgrass
  replacement by *Spartina anglica* 178
  species 172–3
  see also seagrass
effective oxygen load (EOL) 41, *42*
  storm discharge 45
effluent
  discharge 9, 33
    dissolved oxygen 39
  quality from sewage treatment works 44
  see also industrial effluent; sewage
egg predation 83, 123
embankments 20, 54
  materials 51
  natural 65
English Nature 196
Entoprocta species list 209
environment, human impact 11–15
Environment Agency (EA) 21
  flood defences 25–6
  half-tide correction model 68
  invertebrate studies 86
  operating agreements 25
  turbidity data 36
  water quality objectives/standards *24*
environmental impact assessment 196
environmental improvement indicators 198
environmental management, current approach 21, **22**, 23, *24*, 25–6
Eriocheir sinensis 2, 93
estuaries 191–3
  biological criteria for water management 15
  chemistry 194–5
  classification 193
  definitions 194
  degradation 197
  freshwater inflow 192
  human activities 192–3
  pollution 195
  protection 195
  research 194
  sea level rise 196–7
  vulnerability 193, 195
Estuarine and Brackish-water Association 3
estuarine organisms before 1800 10–11
EU Habitats Directive (1992) 196
European Dangerous Substances Directive 46
euryhaline reach 16, 23, *24*, 25
Eurytemora affinis 72–3
  abundance 81
  anadromous migration 77
  diet 81
  epizoic ciliates 84
  faecal pellets 81, **82**
  feeding efficiency 82
  lateral distribution 77, **80**
  predation 83
  vertical distribution 77, **78**, *79*

fibre occurrence 82, 84
fish
  community 138
  diversity before 1800 10
  estuarine 115, 119, 120
  freshwater 119, 120
  importance of tributaries 21
  juveniles 120, 123
  marine 115
    estuarine-dependent 119, 120, 124
    straggler 119, 120
  migration 123
  migratory 119, 120
    parasitic infections 141–3
    water quality 16, 18
  parasite vector 88–9
  seasonal movements 120
  species
    abundance 122–9
    as biological tools 138
    identification 134
  tolerance for estuarine life 115
  trawling 132, 134–5, **136–7**, 138
  tributaries 26
fish-kills 23, 192
  storm drainage 45, 120–1
fisheries
  coarse 23
  commercial 9, 10–11, 115
flatfish 115
Fleet river 192
flood defences 7–8, 21, **22**, 193
  coastline sinking 172
  habitat loss 20–1
  maintenance/replacement 25–6
flounder **119**, 122–4
  abundance 132
  acanthocephalan parasites 163
  condition factor (*K*) 155–6
  copepod ectoparasites 152, 164
  dendrogram 132
  egg predation 123
  G. zaddachi consumption 150
  migration 123, 150
    parasite loss 157
    pattern 166
    spawning 151
  mortality 123
  nematode parasites 142
  P. laevis 143
    colouration 155
    eggs 154
    infection 149, 150–2
    parasite burden 155
    pathogenicity 155
    prevalence **151**, 152
  parasite groups 162–4

flounder *continued*
  spawning 150, 151
  tetraphyllidean parasite larvae 162–3
  trawling 135, **136–7**
flowmeter, fish trawling 134
flukes, parasitic 141, 142
Framework Convention on Climate Change (FCCC) 197
freshwater flow 30–3
  abstraction 32–3
  chloride concentrations 39
  inflow to estuaries 192
  low 113
  temperature 34
  variation 32
  water quality 31–2
freshwater reach 16, 23, *24*
  algae 54–5
Fucus upriver migration 57–8, 63

Gadidae 129, 132
gammarids
  distribution 81
  predation 83
Gammarus pulex 166
Gammarus zaddachi 73, **74, 90**
  benthic macroinvertebrate zone 1 88–90
  distribution 146
  fish parasite vector 88–9
  fish predation 146
  P. laevis infection 166
    cystacanth 150
    patterns 145–8
    shelled acanthor ingestion 144
    zone 146
  predation 88
  sexual dimorphism 147–8
  vertical distribution 77, **78**, *79*
gastropods
  benthic macroinvertebrate zone 1 90, 93
  benthic macroinvertebrate zone 2 94, 97
  benthic macroinvertebrate zone 4 104
Gastrotricha species list 209
geese 174–5
general salt marsh (GSM) community 176
glasswort 175–6
global warming 196
glycophytes 188
Gobiidae 129
goby, sand **119**, 126, 132
granite 110
greater Thames saltmarsh community composition 184–9
  lower marsh 187
  middle-marsh vegetation 188
  Norfolk saltmarsh comparison 185, **186**
  rainfall 188
  sea aster 187
  south coast saltmarsh comparison 184–5
greenhouse effect 196–7

habitats 9
  algal flora 53–5
  elimination 193
  loss
    flood defence work 20–1
    land-claim 20–1
  management 25–6
  modification 193
  restoration 197
Halimione 184
halophytes 175, 188
harpacticoid copepods 71
helminth parasites 141
  eel 164, *165*
Her Majesty's Inspector of Pollution (HMIP) 21
herpes virus
  particles 158, **159**
  smelt infection *163*
  spawning papillomatosis of smelt 158, **159**, 166
herring 126–7
  abundance 132
  trawling 135, **136–7**, 138
hexachlorocyclohexane 47
Hirudinea 91, 104
human activity, environmental threat 5
human impact on environment 11–15
hydrogen sulphide 12, 14
hydroids 110–11

industrial development, habitat loss 20
industrial effluent 191
  controls 20
  discharge 33
Ingrebourne river 21, **22**
insects
  benthic macroinvertebrate zone 2 97
  salinity tolerance 92
Integrated Coastal Area Management programme (UNESCO) 196
Intergovernmental Oceanographic Commission (UNESCO) 196
Intergovernmental Panel on Climate Change 197
intertidal area
  land claim 177, 193
  mudflats 98, 99
Inula crithmoides 187–8

jetties 50
  see also piers

Kinorhyncha species list 209

land-claim 193
  habitat loss 20–1
leeches 91, 104
Lepeophtheirus pectoralis 142
Lernaeocera branchialis 164

Limanda limanda 132
Limonium–Armeria association 184, 185
lindane *see* hexachlorocyclohexane
London County Council 12
  water quality surveys 15
Lowestoft Frame Trawl 134
Lutra lutra 198

macroalgae
  benthic 49
    distribution 58–9, *60–1*, 62–4
  distribution database 52
  freshwater tidal reaches 54
  seasonality 57
macroflora, benthic 52
macroinvertebrate communities, benthic 85–8
  freshwater 88–93
  multivariate analysis 87, **88**
  ordination techniques 87, **88**
  saline tolerance 92
  similarity matrix **87**
  upstream migration in drought 92
  zone 1 88–93
    core community **89**, 91–2
    flow decrease effects 92
    salinity 92, **93**
    soft sediment 93
  zone 2 94–7
    freshwater flows 94
  zone 3 97–100, *101*
    intertidal regions 98–9
    subtidal regions 99–100, **101**
  zone 4 101, **102**, 103–5, **106**, 107–10
    abundance 103
    intertidal zone 101, **102**, 103
    mud stability 109
    north shore 101, **102**, 103–5
    sediment stability 107
    shipping channel 105, **106**
    subtidal area 107–8
    transect 105, **106**, 107–10
  zone 5 110–12
    diversity at sewage outfall 112
    substrate 10
macroinvertebrate distribution 85
Marine Conservation Society 196
marine environment 191
marine reaches 16, *24*
  algal flora 53–4
marsh
  grazing 9, 20
  loss 51
  see also mudflats; saltmarsh
Medway river 7
Merlangius merlangus 124–5
metals
  contamination 191
  sediment concentrations 47
Metropolitan Board of Works 12
  interceptor sewers 20
microalgae 55
microflora, benthic 52
Microstomus kitt 132
middle-marsh vegetation of greater Thames saltmarsh 171, 176, 188
midge, phantom 97
migration
  anadromous 77
  fish 16, 18, 119, 120, 123, 141–3
  flounder 123, 150, 151, 157, 166
  human population to coast 193
  salmon/salmonids 14, 16, 18, 23
  upriver of *Fucus* and *Ulva* 57–8, 63
  vertical of zooplankton 77, **78**, *79*
Minimata incident 191
mitten crab, Chinese 2, 93
modelling 47–8
Mollusca
  benthic macroinvertebrate zone 1 90
  benthic macroinvertebrate zone 4 103, 105, 108
  species list 226–8
Monogenea 141
  eel parasites *165*
mud reaches 36
mudbanks
  algae 55
  benthic macroinvertebrate zone 4 105
mudflats
  benthic macroinvertebrate zone 2 94, 95
  colonization by vegetation 170
  intertidal 20, 94, 169
  productivity 169
  Spartina x townsendii planting 177
  stabilization 170, 177
  subtidal 94
  see also Zostera beds
mussel beds 51
mysids 71
  distribution 81
  predation 83

National Nature Reserves (NNR) 25
National Rivers Authority (NRA) 21, 86
Natura 2000 196
Nature Reserve Agreement 25
Nematoda 141
  benthic macroinvertebrate zone 3 100
  eel parasites *165*
  life cycle 141–2
  smelt 162, *163*
  species list 210–16
Nematomorpha, species list 209
Nemertea
  benthic macroinvertebrate zone 4 104
  species list 208
Neomysis integer 73–4, **75**
  predator–prey relationships 83
nitrate concentrations 45–6
nitrification, water temperature 35

non-biodegradable substances
  control 20
  discharge 15
nuclear weapons testing 191
nutrients 45–6
  output 192

oligochaetes
  benthic macroinvertebrate zone 1 91, 93
  benthic macroinvertebrate zone 2 94, 95, 96–7
  benthic macroinvertebrate zone 3 98, 99, 100
  benthic macroinvertebrate zone 4 103, 104, 107, 108, 109
  freshwater 94
ophiuroids 108
opisthobranchs 104
orache 175
orthophosphate 46
Osmeridae 129
Osmerus eperlanus 127–8, 198
  Pomphorhynchus laevis infection 149–50
  virus-associated spawning papillomatosis 157–62
otter 198
oxygen
  balance and water temperature 35
  depletion in twentieth century 13–14
  solubility 35
  see also biochemical oxygen demand (BOD); effective oxygen load (EOL)
oxygen, dissolved 39, 41–5
  concentration 16
  effluent discharge 39
  pollution budget 18
  reduced flow impact 43–4
  standard 23
  summer storms 44–5
  supersaturation 42
  variation with time 41–2
oxygenation, in-river unit 19, 45

paper mills 33, 82
papillomatosis *see* spawning papillomatosis of smelt
Paracapillaria 163
parasites
  estuarine fauna 141
  host species 142
  life cycle strategies 141–2
  microhabitat specificity 142
PAUP 62, 63
pebble community 95, **96**
pesticides, organohalogen 20
Phaeophyta
  distribution 58
  local variation 62
phosphate concentrations 45, 46
physicochemical parameters *200*
phytoplankton 42, 50
piers
  floating 54
    algal assemblages 62
    pioneer colonization 58
  substrate for benthic organism attachment 50
pilings, metal 54
Pippard Committee 14–15
  public nuisance 16
plaice
  dendrogram 132
  trawling 135, **136–7**, 138
plankton, pollution 68
plant material, decaying 108
Platichthys flesus 122–4, 149
Platyhelminthes
  benthic macroinvertebrate zone 1 91–2
  parasitic 141
  species list 208
Pleuronectes platessa 132
Pleuronectidae 129
pogge 138
pollution 191
  budget 15
    dissolved oxygen 18
  estuaries 195
  improvements 113
  industrial 11–12
  plankton 68
  Port of London Authority 15
  precautionary control regimes 196
  sewage 192
  tolerance 23
  water quality 196
polychaetes
  benthic macroinvertebrate zone 2 97
  benthic macroinvertebrate zone 3 98, 100
  benthic macroinvertebrate zone 4 103, 104, 105, 107, 108
  benthic macroinvertebrate zone 5 111
Pomphorhynchus laevis 143–4, **145**
  colouration 155
  cystacanth 144, 146, 147
  definitive host species range 148–50
  eel infection 164
  estuarine strain 165
  extra-intestinal location 144
  female host size 147–8
  flounder
    condition factor (*K*) 155–6
    host 149
    infection patterns 150–2
    loss from during migration 157
  G. zaddachi
    infection pattern 145–8
    intermediate host 166
  host
    age-related prevalence **151**, 152
    distribution 166

immune response 154–5
intermediate 144, 145–8
utilization patterns 148–9
intestinal attachment 144
life cycle **144**
indirect 166
parasite burden 155–7
pathogenicity 155–7
proboscis bulb 143–4, **145**
seasonal trends 146, **147**
shelled acanthors 144, 153
smelt infection 152–5
strains 143
transmission to fish 146
water temperature 146, **147**
port 9
Port of London Act (1964) 15
Port of London Authority 15
Port of London (Consolidation) Act (1920) 15
post-parasite interactions 141
Potamopyrgus jenkinsi 90, **91**
Pomatochistus minutus 126
pouting 132
power generation, tidal 193
power stations 8–9
cooling water 8–9, 116–17, 118
aeration 117
discharge 34
warm water
discharge 13–14
effluent 23
see also West Thurrock Power Station
prawn 100
predator–prey relationships 83
Protozoa 69
species list 206–7
public health 192
Puccinellia maritima 179, 184, 185
greater Thames saltmarsh 186
Puccinellia–Festuca grass community 185

RAMSAR Convention 196
RAMSAR site 25
recreational facilities 193
residential development 20
retention time 30
Rhodophyta
distribution 58
local variation 62
Richmond
half-tide barrier 7, 10
high-tide lock 30
river systems, integrated management 195
river wall
algal zonation 54
see also embankments
rockling, four-bearded 132
rocky shore 9
Rotifera 70
species list 209
Royal Commission on Environmental Pollution (1972) 15
RSPB 196

sailing 9
Salicornia 175–6
greater Thames saltmarsh 186
replacement by *Spartina anglica* 178
saline tolerance of macroinvertebrates 92
salinity 37, **38**, 39
benthic macroinvertebrate zone 1 92, **93**
benthic macroinvertebrate zone 2 94
benthic macroinvertebrate zone 3 97
Salmo salar 128–9, 198
salmon
Atlantic 128–9
migration 23
re-establishment 198
salmonids 10
migratory 14, 16, 18
salt-pans 188
saltmarsh 9, 51
accretion 170, 172
algae 54
classification 176, 180, **181**
cliff erosion 171–2
community types *182*
definition 170
development 170–2
dichotomous key 180
east coast 176, **177**
English Nature survey 183–5
erosion 170–1
expansion/contraction cycles 170–1
greater Thames community composition 184–9
middle-marsh community 176, 188
loss 171
nodum groups 180–2
NVC class *182*
organic matter production 169
plant communities 169–70
profile **171**, 172
regional communities 179–85, **186**
restoration 197
seaward extension 170
Spartina anglica invasion 176–9
tidal inundation 170
Types A, B and C 180–1
vascular plants 175–6
vegetation
determining factors 189
noda 179–80
productivity 169
zonation 170
wildfowl grazing 189
see also greater Thames saltmarsh community composition
samphire, golden 187–8
sand banks 93

sand dunes 9
sand goby **119**, 126
  abundance 132
sandflats 104
sea aster 187, 188
sea defences *see* flood defences
sea level rise 172, 196–7
sea-blite 187
sea-lavender–thrift association 184, 185
sea-purslane 184, 185
  pedunculate 188
sea-walls
  algal flora 53
  erosion acceleration 172
  materials 51
  see also flood defences
seagrass
  angiosperm species 172
  communities 169–70
  decomposing mats 108
  intertidal communities 172–5
sediment
  accumulation 37
  anoxic 35
  mats 55
  metal concentrations 47
  suspended **82**
Sertularia compressina 110–11
settlement surfaces
  artifical 55
  seasonal species richness 56–7
sewage
  chemical treatment 12
  discharge 191–2
    phosphate 46
  effluents 8
  pollution 192
sewage treatment
  nitrate levels 45–6
  water quality improvement 19
sewage treatment works **6**, 8, 13
  effluent quality 44
  freshwater flow 31
  improvements 14–15, 19
  outfall 113
  secondary treatment plants 13
sewers, interceptor 12, 20
Shannon–Wiener diversity index 121–2
shingle banks 51
shipping 9
Shoeburyness 6
silt mat stabilization 55
similarity matrix **87**
Sipuncula species list 209
Sites of Special Scientific Interest (SSSI) 25
smelt **119**, 127–8
  condition factor (*K*) **157**
    analysis 161
  Cryptocotyle lingua intermediate host 162
  digenean parasites 162
  G. zaddachi consumption 150
  nematode parasite 162, *163*
  P. laevis
    colouration 155
    eggs 153–4
    host immune response 154–5
    infection 149–50
    infection patterns 152–5
  return to Thames 198
  spawning papillomatosis 166
  trawling 135, **136–7**, 138
  tumours 158, **159**
  virus-associated spawning
      papillomatosis 157–62
soft engineering options 26
sole 124
  dendrogram 132
  Dover 198
  lemon 132
  trawling 135, **136–7**, 138
Solea solea 124, 198
Spartina anglica 175, 189
  colonization 177
  die-back 179
  greater Thames saltmarsh 186
  long-term effects 179
  productivity 178
  saltmarsh invasion 176–9
  seaward development 179
  spread 178
  tidal submergence tolerance 178
Spartina maritima 187
Spartina x townsendii 177–8
spawning papillomatosis of smelt 157–62,
    166
  gonadal maturation 160
  lesion bacterial colonization 161
  morphology 158
  seasonal prevalence 158–9
  spawning time 161
  tumour prevalence **162**
  ultrastructure 158
Special Areas of Conservation (SACs) 196
Special Protection Areas (SPAs) 25, 196
species numbers, site records **98**
Statutory Quality Objectives (SQOs) 196
storm discharge
  dissolved oxygen 23, 44–5
  EOL 45
  tributary opening 26
  water quality 19–20
storm drainage
  fish-kills 120–1
  untreated sewage 113
storm surge control 193
stressors 84
Suaeda maritima 187
supersaturation, dissolved oxygen 42
surfactants 13
suspended solids

concentration profiles **36**
tidal cycles 36–7
tides 35
variations 35–6
see also turbidity

tapeworms 141
tetraphyllidean 163
Tardigrada species list 209–10
Teddington Weir 6, 7, 10
flow 32, **33**
freshwater flow 92
storm discharge management 20
temperature *see* water temperature
tetraphyllidean parasites 162–3
Thames Barrier 7, 21
Thames Bubbler 19, 45
Thames estuary
algae sampling sites **52**
before 1800 9–11
depth 7–8
ecosystem rehabilitation 19
freshwater inputs 8
greater 6–7
historical accounts 9–10
reach divisions 21
region **1**
study area 6–7
use 8–9
width 7–8
Thames Estuary Benthic Programme (TEBP) 86
Thames river 7
see also tributaries
Thames Salmon Rehabilitation Scheme 128–9
Thames Water Authority
EA agreements 25
non-biodegradable substance discharge 20
toxic substance discharge 20
water quality
management 16, *17*, 18–20
objectives 16, *17*, 18
Thames Water Utilities 32
tidal excursion 67
zooplankton sampling 77
tidal flow, chloride concentrations 37, 39
tidal inundation of saltmarsh 170
tidal levels 27
tidal power generation 193
tidal range 67
tides
neap 28
oscillations 28
spring 28
suspended solids 35
water levels 28
water movements 30
toilet tissue breakdown 82
topography 27, **28**
toxic substance discharge 15, 20
trace metals 35
trawling 132, 134–5, **136–7**, 138
tributaries 8, 10, 192
closing off 21, **22**
fish 21, 26
freshwater flow *31*
incorporation into sewerage system 21, 192
re-opening mouths 25–6
Trisopterus 132
tubificid worms 1–2
turbidity 35–7
light penetration 51
TWINSPAN 52, 62, 63, 64

UK Red List 46
Ulva upriver migration 57–8
Uniramia species list 224–6
urban renewal 193

vector, fish parasite 88–9

wading birds 189
Waste Regulation Authorities 21
water
demand for 32
deoxygenation 192
levels 28–30
movements 30
velocity 30
viscosity 35
Water Act (1989) 196
Water Pollution Research Laboratory study 14–15
water quality 11
automatic monitoring stations 19
biological standards 23, *24*, 25
decline
nineteenth century 11–12
twentieth century 13–14
fish species as biological tools for monitoring 138
flow 31–2, 33
management 21, 23, *24*, 25
Thames Water Authority era 16, *17*, 18–20
objectives *17*
Pippard Committee recommendations 14–15, 16
sediment resuspension 35
sewage treatment improvement 19
standards *17*
storm discharge 19–20
surveys 10
London County council 15
see also pollution

water temperature 33–5
  atmospheric conditions 34
  estuary processes 35
  Pomphorhynchus laevis 146, **147**
  standard 23
  variations 34
West Thurrock Power Station fish studies 116
  abundance **119**
  categories 119
  community comparison 129, *130–1*, 132
  cooling water demand 118
  diversity 121–2
  fish measurement 118
  fish-kills 120–1
  juveniles 120
  sampling methods 116–18
  screens on water intake 117
  seasonal movements 120
  species
    abundance 122–9, *130–1*
    dendrogram 132, **133**
    diversity 118
    increases 122
    numbers 118–21
    per sample 120, **121**
  water extraction 116–17
whiting **119**, 124–5, 132
  copepod parasites 164
  trawling 135, **136–7**, 138
widgeon 174
wildfowl 174, 189
Wildlife and Countryside Act (1981) 25
windsurfing 9

zinc 47
zooplankton
  allochthonous 67
  feeding 81–2
  food organisms 83
  functioning 81–4
  predator–prey relationships 83
  taxa 69
zooplankton communities 67–8
  classification scheme 68
  comparison with other estuaries 80–1
  composition 69–72
  dominance changes of major species 74, *76*, *77*
  lateral distribution 77, 80
  particulates as stressors 84
  salinity range 68
  sampling zones **69**
  seasonal distribution 74, *76*, *77*
  spatial distribution 72–4
  study 68, *70*
  temporal distribution 72–4
  vertical distribution 77, **78**, *79*, 80
Zostera 169
Zostera beds
  Brent goose grazing 174–5
  detritus food chain 175
  die-back 172
  greater Thames 189
    saltmarsh community composition 186
  macrofauna 173
  meiofauna 173
  pioneer 174
  plant groups 173
  Spartina anglica invasion 175, 178
  wasting disease 175, 178
  widgeon 174
Zostera marina 172, 173, 174–5
  intertidal 173
  wasting disease 175
Zostera noltii 172, 173, 174

MIX
Papier aus verantwortungsvollen Quellen
Paper from responsible sources
FSC® C105338

If you have any concerns about our products,
you can contact us on
ProductSafety@springernature.com

In case Publisher is established outside the EU,
the EU authorized representative is:
Springer Nature Customer Service Center GmbH
Europaplatz 3, 69115 Heidelberg, Germany

Printed by Libri Plureos GmbH
in Hamburg, Germany